Proceedings of the Royal Institution of Great Britain

Proceedings of the Royal Institution of Great Britain

Volume 67

Edited by
P. DAY

Oxford New York Tokyo
OXFORD UNIVERSITY PRESS
THE ROYAL INSTITUTION
1996

Oxford University Press, Walton Street, Oxford OX2 6DR
Oxford   New York
Athens   Auckland   Bangkok   Bombay
Calcutta   Cape Town   Dar es Salaam   Delhi
Florence   Hong Kong   Istanbul   Karachi
Kuala Lumpur   Madras   Madrid   Melbourne
Mexico City   Nairobi   Paris   Singapore
Taipei   Tokyo   Toronto
and associated companies in
Berlin   Ibadan

Oxford is a trade mark of Oxford University Press

Published in the United States
by Oxford University Press Inc., New York

A catalogue record for this book is available from the British Library

Library of Congress Cataloging in Publication Data
Data available

ISBN 0 19 855938 0

Typeset by Footnote Graphics, Warminster, Wiltshire
Printed in Great Britain by
Redwood Books, Trowbridge, Wiltshire

PREFACE

For over a century, one of the services that the Royal Institution has provided to its Members has been to furnish them with an annual volume containing the texts of the Institution's famous Friday Evening Discourses. As such, the resulting volumes of what we call the Proceedings must represent the longest continuously sustained contribution to enhancing public awareness of science and technology anywhere in the world. It is therefore with the greatest pleasure that we bring to you the sixty-seventh volume in the present series, containing accounts of topics drawn from the widest range of disciplines, from astrophysics to biology and artistic perception. The kaleidoscope of science continues ever brighter and colourful, impinging on more and more aspects of our lives. The topics in this volume give ample evidence of that vigour and variety.

London   P. D.
July 1996

CONTENTS

PLATES

1. Photograph taken by Robin Whyman of the Geodesic Dome designed by Buckminster Fuller for the US exhibit at Montreal EXPO67. One of the pentagons necessary for closure is discernible in this beautiful photograph.

2. The two card models which played key roles in the positing of the truncated icosahedral structure of C_{60} (see text). (a) Stardome map of the sky (Buckminster Fuller patented t-icosahedral, and other polyhedral, world map projections); (b) the prototype model constructed by Smalley. Both models are truncated icosahedra with 60 vertices, 12 pentagonal and 20 hexagonal faces.

3. Model structure of the giant fullerene C_{240}.

4. Original 'red(d)ish' extract obtained on Monday 6th September 1990.

5. Chromatographic separation by Taylor produced the magenta solution of C_{60} (left) and red solution of C_{70} (right).

6. Computer simulation of the condensation of gas clouds to form a star cluster (*after reference 1*). The red regions represent high densities; the image shows two protostellar clusters in orbit around each other.

7. Model of the circulation of ocean currents around the Antarctic continent.

8. Computer model of the distribution of pressure over a fighter aircraft in flight. Blue colours indicate regions of low pressure; red colours show high pressure regions. (Kindly supplied by CRAY Research (UK) Ltd.)

9. The molecular structure of the enzyme lysozyme with a sugar molecule bound at the active site.

10. The crystal structure of the superconductor lanthanum copper oxide (La_2CuO_4); red spheres are oxygen atoms, blue spheres are lanthanum and yellow spheres are copper.

11. Schematic representation of the connectivity of the CRAY T3D massively parallel supercomputer. The right hand side represents the interconnected processors.

12. The crystal structure of Li_3RuO_4 generated from an initial random arrangement of atoms. The purple polyhedra represent RuO_6 octahedra and the turquoise polyhedra show the coordination spheres of the lithium ions.

hand the density of gas. The three states shown are separated in time by approximately 1.5 billion years. (Kindly supplied by Prof. A. Nelson.)

28. Ribbon diagram of the IL2 complex. The four long helices of IL2 are in red, while the short helix of the AB loop is magenta and the short β-strands pink. The α chain is yellow and white, the β chain blue and cyan, and the γ_c chain green. The surface of the area buried between IL2 and the β and γ_c chains is shown by white stippling. (Photograph courtesy of Dr Graham Richards.)

29. Young male with xeroderma pigmentosum; basal cell carcinoma, squamous carcinoma and melanomas have all occurred on the head and neck.

30. Allergic dermatitis to the nickel in a zip on a pair of jeans. This type of reaction is very difficult to predict and in these rarer forms, is seen at a very low frequency in the population. Thus, penicillin, a most valuable drug, is nevertheless responsible for a number of deaths each year from allergic reactions whilst being enormously valuable for the population as a whole.

31. A picture, used for teaching doctors how to remove IUDs, which uses the same convention—not like the 'warm reality' at all (from Camera Talks).

32. Another slide from the same series as above, showing the Dalkon Shield 'in place'; it is clearly designed not to fall out of the icon. It caused much pathology when used in real women (from Camera Talks).

33. The Belousov–Zabotinski reaction: a recursive chemical reaction, demonstrating that complexity can arise from simplicity, that the complexity of a human being can result from interactions during development and need not be laid out in the DNA.

34. DNA cannot 'make' organisms. (Reprinted courtesy *OMNI Magazine*, copyright 1988.)

35. A computer-generated model of the deformation behaviour of a rubber-toughened epoxy adhesive. The finite-element analysis shows two (one-quarter) spherical rubber particles. The material between the rubber particles is the epoxy adhesive. The green-coloured band represents the development of a plastic shear band in the epoxy polymer—initiated by, and running between, the rubber particles. (The adhesive is being loaded vertically.)

36. Glow worms releasing their own light in the dark.

37. A ten year old's recollection of his first encounter with chemistry.

CONTRIBUTORS

Akira Tonomura
Advanced Research Laboratory,
Hitachi Ltd., Hatoyama,
Saitama 350−03,
Japan

Harold Kroto
School of Chemistry and
    Molecular Sciences,
University of Sussex,
Brighton BN1 9QJ

Sir Martin Rees
King's College,
Cambridge

C. Richard A. Catlow
Davy Faraday Research
    Laboratory,
The Royal Institution of Great
    Britain,
21 Albemarle Street,
London W1X 4BS

Joan Bordas
Department of Physics,
University of Liverpool,
Oliver Lodge Laboratory,
Oxford Street,
Liverpool L69 3BX

Sir Colin Berry
Department of Morbid Anatomy,
The Royal London Hospital,
London E1 1BB

Lewis Wolpert
University College and Middlesex
    School of Medicine,
Department of Anatomy and
    Developmental Biology,
Windeyer Building,
Cleveland Street,
London W1P 6DB

Tony Benn, MP
House of Commons,
London SW1

Jack Cohen
Institute of Mathematics and
    Ecosystems Analysis and
    Management Unit,
Department of Biological
    Sciences,
University of Warwick,
Coventry CV4 7AL

Anthony J. Kinloch
Imperial College of Science,
    Technology and Medicine,
Department of Mechanical
    Engineering,
Exhibition Road,
London SW7 2BX

Ann Hubbard
Reigate College,
Castlefield Road, Reigate,
Surrey RH2 0SD

Raymond Keene OBE
86 Clapham Common Northside,
London SW4 9SE

Allan Chapman
Wadham College,
Oxford OX1 3PN

Rosalyn Tureck
Tureck Bach Research Foundation,
Windrush House,
Davenant Road,
Oxford OX2 8BX

Electron waves unveil the microcosmos

AKIRA TONOMURA

Light is a medium which conveys to us information about outside worlds ranging from the microscopic world through microscopes to the universe through telescopes. However, there is a fundamental limit on the size of an object that can be seen with light. The limit is given by the wavelength of the light used; since the wavelength of light is on the order of 0.5 μm, an object smaller than that cannot be seen, even through a microscope.

Have we no means of seeing the world of objects smaller than the wavelength of light?—Yes, we can do so by using waves of shorter wavelengths. Among several candidates such as X-rays and neutrons, electrons accelerated up to 100 kV have wavelengths as short as 0.03 Å, and microscopic objects can be seen through electron lenses. Since our eyes are not sensitive to electrons, we can observe the image of an object only after transforming the intensity of an electron beam into a light intensity, such as emitted from a fluorescent screen or the gradations of a photograph. This procedure, called electron microscopy, allows an object to be seen on an atomic scale, which is far smaller than the wavelength of light.

However, the techniques of electron optics are still undeveloped compared with the techniques of light optics. To take an example, the phase-contrast microscopy invented by Zernike [1] made optical microscopy come to life again by making living bodies in motion visible without killing them by staining. Such techniques are not yet available in electron microscopy. This is because there are no simple electron counterparts of convenient optical parts such as mirrors and half-mirrors. There is a magnetic convex lens but no concave lens. There is no coherent electron beam corresponding to a laser beam.

This discourse introduces a new kind of electron microscopy that makes use of the phase information of an electron wave. This field has just been opened up by the development of a 'coherent' field-emission electron beam [2] and electron holography [3]. The coherent electron beam is indispensable for electron holography; through electron holography the electron image is transformed into an optical image, thus allowing use of versatile optical techniques in electron optics.

Electron waves

Electrons are too small to see, so the reality of electrons has been an object of discussions for more than 100 years. J. J. Thomson discovered in the late 1800s that electrons are elementary particles.

However, electrons were demonstrated to behave as waves as well by Davisson and Germer [4], G. P. Thomson [5] and Kikuchi [6]. This apparent contradiction was consistently explained by quantum mechanics [7], though it is too strange to understand intuitively; electrons can behave as waves as long as they are not detected. When detected, however, they are always single particles. This strange explanation has often been challenged theoretically, for example, by Einstein [8], and Bohm [9], and tested in experiments by Aspect [10], but no trials up to now have succeeded in negating quantum mechanics.

On the contrary, the strange behaviour of electrons was caught as a movie in our experiments [11]. Electrons emitted from a source pass through an electron interferometer called an 'electron biprism' [12] and are detected on the monitor of a two-dimensional detector (Fig. 1). Electrons are detected one by one as bright spots at the corresponding locations of the electron arrivals on the monitor (see Fig. 2(a)). So long as electrons are detected as points, we see that they are particles, consequently, they must have passed through on one side of the central filament of the electron biprism or the other side.

Electrons are emitted from the source one after another at such a long time interval in the present experiment that the average distance between consecutive electrons is much larger than 100 m. Since the apparatus is only 2 m long, only one electron exists in the microscope at any instant.

Since an electron has never been found to be divided into fractions, you might think that the wave nature of electrons will never show up in this experiment. In fact, electrons arrive seemingly at random positions, as shown in Fig. 2(b). However, this is not the case: electrons can be seen to behave like waves as well.

When a larger number of electrons have been accumulated, something

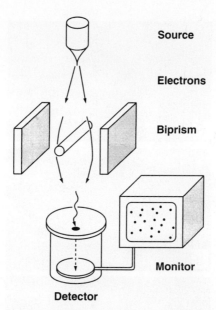

Fig. 1 Schematic diagram of an experimental set-up to demonstrate the formation of an electron interference pattern. Electrons passing through an electron biprism are detected one by one by a two-dimensional detector.

like interference fringes begin to appear, as can be seen in Fig. 2(c). You may recognize the fringes in the vertical direction.

The interference fringes are clearly seen in Fig. 2(d) and (e). These fringes are nothing but those formed as a result of interference between two waves that have passed through on both sides of the filament of the electron biprism. Consequently, we have to accept the very strange conclusion that electrons are detected one by one as particles, but that the whole ensemble manifests the wave property of forming an interference pattern. Quantum mechanics tells us that we have to give up the reality of the particle picture of electrons, except at the instant they are detected.

In the above experiments, it took more than 20 minutes for the interference fringes to appear. We cannot wait so long if we are to use the wave nature of electrons to observe the microcosmos, especially when we want to observe its dynamics.

What is the minimum time needed to form one interference pattern on film? The minimum time needed for obtaining high-contrast interference fringes is a few minutes when conventional thermal electrons are used. We could never see such fringes on a fluorescent screen with the naked eye, but only on film after long exposures. The situation changed when we developed a 'coherent' field-emission electron beam in 1979 [2].

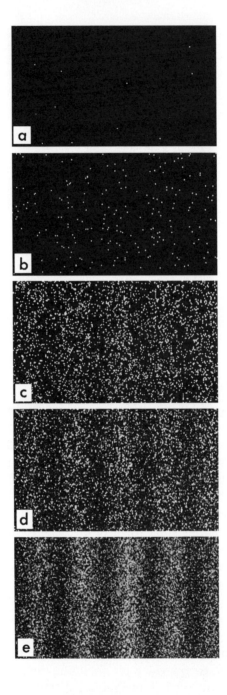

Fig. 2 Single-electron build-up of an interference pattern. The electron density is so sparse that only one electron exists at one time in the apparatus. However, when electrons are accumulated, the biprism interference pattern, which is formed between two electron waves passing through on both sides of the biprism, can be observed. It is as if a single electron splits into two, and passes through on both sides of the biprism: (a) total number of electrons 10; (b) total number of electrons 100; (c) total number of electrons 3000;(d) total number of electrons 20 000; (e) total number of electrons 70 000.

Coherent electron beam

Thermal electrons can be extracted from a heated metal by making electrons inside the metal climb over the surface potential barrier by adding thermal energy. Instead of these electrons, we can use those emitted by a high electric field. The principle is the same as that of a lightning rod; as a thundercloud approaches, many lines of electric force from the cloud reach the lightning rod to concentrate at the sharpened tip, thus generating a very intense electric field. A thunderbolt therefore strikes at the tip.

If we apply a high voltage to the sharp tungsten tip whose electron micrograph is shown in Fig. 3, it can attract a lot of lines of electric force. This results in the emission of electrons from the tip, if not a thunderbolt. The word 'field emission' is used because the intense electric field extracts electrons from the sharp end of the tip. To realize electron emission, we have to sharpen the tip to a diameter of approximately 0.1 μm, so that the electric field near the tip is as high as a few volts across 1 Å. We can make such a tip by electrochemical polishing. The electrons come out of the sharp end of this tip so that the tip can generate a large number of electrons per unit surface and per unit time. In other words, the field emission gun is a very bright point source of electrons. In addition, the velocity of electrons is very uniform, since there is no need to heat the cathode. As you know, heat has a randomizing effect that results in non-uniform velocities of electrons. These features allow us to obtain highly coherent electron waves from the field-emission gun. With this electron beam, the number of interference fringes recordable on film increased from 300 to 3000 (Fig. 4), clear evidence of the high coherence. Moreover, owing to the high brightness of the electron beam, we can see interference fringes directly on a fluorescent screen with our naked eye, exactly as we can do with optical lasers. Because of the high coherence of this electron beam, we can make Gabor's dream come true, achieving electron holography [13].

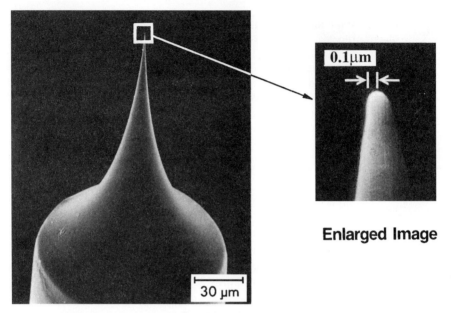

Enlarged Image

Fig. 3 Scanning electron micrograph of a field-emission tip. Electrons are emitted from the tip 1000 Å in radius owing to the tunnelling effect caused by a high electric field at the tip surface. Compared with conventional thermal electrons, the electric current density is higher by five orders of magnitude, the virtual source size of 50 Å is smaller by three orders of magnitude, and the energy spread is narrower by a factor of three. Because of these features, an electron interference pattern is observable on a fluorescent screen with the naked eye.

Electron holography

Electron holography is a two-step imaging technique (Fig. 5). In the first step, an electron wave that has passed through an object is superposed on an undisturbed electron wave. The resulting interference pattern is recorded on film as a 'hologram'. When this hologram is illuminated by a laser beam, the wavefronts are optically reconstructed; in other words, the object image is reproduced in three dimensions. Once the electron wavefronts are reproduced as light wavefronts in this way versatile optical techniques can be introduced.

In an ordinary electron microscope we can observe only the intensity of the electron wave that has passed through an object, that is to say, the silhouette. In electron holography, however, we can also read the phase of the electron wave by optical techniques. The reconstructed 'phase' image is called an 'interference micrograph'.

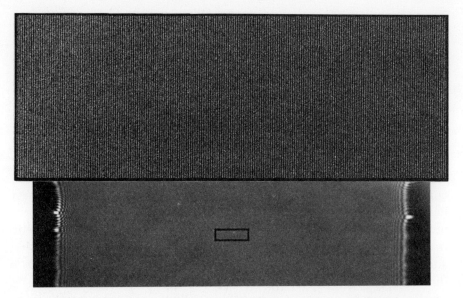

Fig. 4 Electron interference fringes formed with the electron biprism. Owing to the high coherence of the field-emission electron beam, the number of interference fringes recordable on film has increased from 300 to 3000. Only a part of the interference pattern (*below*) is enlarged to show the fringes (*above*).

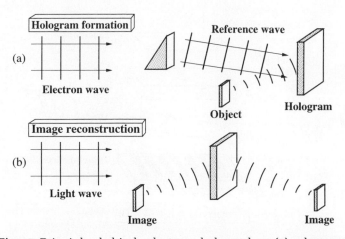

Fig. 5 Principle behind electron holography: (a) electron hologram formation; (b) optical image reconstruction. Electron holography is a two-step imaging technique. An object wave is recorded on film as an interference pattern called a *hologram* by overlapping a reference wave. Then illumination by a laser beam forms an optical image of an object. The electron wavefronts are faithfully reconstructed as optical wavefronts.

The phase of an electron wave is independent of the intensity. Consequently, interference micrographs sometimes provide information about an object that is inaccessible by electron microscopy. An example is electromagnetic fields in the microcosmos.

Visualizing lines of magnetic flux

Lines of magnetic force emanating out from a horseshoe magnet can be seen by sprinkling iron filings on a glass plate covering the magnet. The concept of lines of magnetic force was first devised in 1831 by Michael Faraday. Furthermore, he recognized in 1851 that lines of magnetic force must penetrate into the magnet and proposed the concept of magnetic induction, or magnetic flux density, to describe the magnetic state inside and outside the magnet. He noticed that magnetic induction is like a flux, having neither sources nor sinks. The magnetic flux is visualized by the lining up of magnetized iron filings along lines of magnetic flux. While the picture drawn by iron filings is not quantitative, the interference electron micrograph [14] was found to provide a quantitative picture of lines of magnetic flux.

The principle behind such observation can be explained in simple terms. It is illustrated in Fig. 6. A coherent electron beam is incident on a

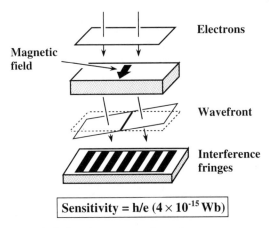

Fig. 6 Principle behind magnetic-flux observation. An incident planar wavefront is rotated round a line of magnetic flux, consequently, contour lines of the transmitted wavefront indicate projected lines of magnetic flux. Such contour lines become directly observable by overlapping a 'dotted' reference plane wave as an electron interference micrograph. This has become possible by electron holography.

uniform magnetic field. The electrons are deflected by the magnetic field in the direction perpendicular to the field. What would you expect to happen to the electron wavefront, or the surface perpendicular to the electron trajectories? The wavefront is tilted or rotated about the axis of the magnetic field. The rotation direction depends on the direction of the magnetic field. In this way, information about the magnetic field is reflected in the shape of the wavefront of the electron wave transmitted through the magnetic field.

To obtain the interference micrograph, let us superpose this rotated wavefront with a reference wavefront, the dotted one in Fig. 6, that has bypassed the magnetic field. The wavefronts of the two waves overlap along the intersecting line, enhancing the intensity owing to interference. Such lines of enhanced intensity are parallel to the lines of magnetic flux, for they are the axes of the wavefront rotations. Thus, the lines of magnetic flux can be observed directly as the contour lines of the wavefront in an interference micrograph.

It can be shown by simple calculation that the amount of magnetic flux flowing between the two adjacent interference fringes is a constant, h/e, where h is Planck's constant and e is the charge of the electron. Therefore, analysis of the interference fringes can give precise, quantitative information about the distribution of lines of magnetic flux.

An example is shown in Fig. 7. The specimen here is a magnetic tape. The information 'yes' and 'no' is recorded on a cobalt thin film as an array of tiny magnets by moving over the film a magnetic head, whose north and south poles are changing alternately. The recorded film is then peeled off and observed.

When observed by electron microscopy, nothing can be seen concerning the recorded information other than the film structure. In the interference micrograph shown in Fig. 7, though, many interference fringes appear. They indicate the lines of magnetic flux inside and outside the film. A constant magnetic flux of h/e is flowing between two adjacent fringes. What is interesting about this picture is that we are looking at the transparent image of a tape, which is a solid object we can touch. At the same time, we are looking at lines of magnetic flux, which we cannot touch. This is an example of what we can do with electron holography.

Faraday predicted that magnetic flux is like a stream of water flow, having neither sources nor sinks. This picture shows that this is indeed the case; at the boundaries where two opposite magnets face each other, lines of magnetic flux collide head on with each other producing vortex-like streams. They meander towards the film edge and leak outside. These regions can be compared with the case where the north poles of two magnets face each other so that the lines of magnetic flux flow

Akira Tonomura

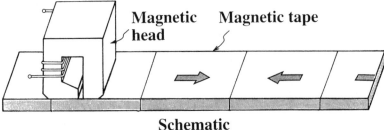

Schematic

Fig. 7 Interference micrograph of recorded magnetic tape. Contour fringes in the micrograph indicate lines of magnetic flux. A constant magnetic flux of h/e is flowing between two adjacent contour lines. A flow of magnetic flux is just like a water flow having neither sources nor sinks. When two opposite lines of magnetic flux collide head-on, vortex-like streams are produced.

outside from the north poles to the neighbouring south poles. The whole space is occupied by lines of magnetic flux.

How densely can we pack these magnets in a tape? Obviously, the distance between the neighbouring magnets cannot be smaller than the width of this vortex region. If the vortex size could be reduced somehow, a higher density recording would be achieved.

Can we make the vortex smaller by changing certain conditions, such as the tape material, the gap of the magnetic head, or the spacing between the head and the tape? Yes, we can! What we obtained when we achieved high density recording is shown in Fig. 8. We can see how densely these magnets are packed after changing the conditions. The

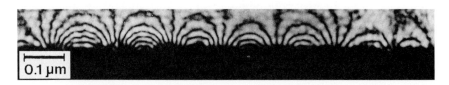

Fig. 8 Interference micrograph of magnetic tape recorded at high density. It is demonstrated that the information 'yes' or 'no' can be recorded in a region as narrow as 0.1 µm.

minimum distance between neighbouring magnets of 0.1 µm is less than one tenth that in Fig. 7.

Applications are not limited to the observation of magnets, but extend to observing superconductors, in which an electric current flows without resistance or energy loss.

Observing superconducting vortices

A magnetic field generally does not penetrate a superconductor; the lines of magnetic flux go around the superconductor. This is called the Meissner effect. Nevertheless, if you make the field sufficiently strong, the lines of magnetic flux do penetrate the superconductor.

Inside the superconductor, the lines of magnetic flux do not spread uniformly, but are tied up in bundles called magnetic vortices. Although there can be many of these vortices in a superconductor when it is placed in a magnetic field, each vortex has a minimum amount of magnetic flux, $h/2e$, so that it is also called a flux quantum.

It has been a dream of electron microscopists for thirty years to observe these vortices directly, but they are too tiny. Although they are so tiny, their behaviour is directly connected to practical applications. Therefore, we applied electron holography to see for the first time the lines of magnetic flux leaking from individual vortices in a superconductor [15] in 1979. In this experiment, an electron wave is incident parallel to the superconductor surface, and the lines of magnetic flux are observed sideways by projection.

The resultant picture is shown in Fig. 9. This picture is amplified by a factor of two in the measurement sensitivity of magnetic flux usin

Fig. 9 Interference micrograph of lines of magnetic flux leaking out of magnetic vortices in lead film 0.2 µm thick (phase amplification × 2). The lower black part indicates the film and the upper fringes indicate lines of magnetic flux. One fringe corresponds to a single vortex. In addition to isolated vortices in the right hand side, an antiparallel pair of vortices was found.

a technique peculiar to holography [3], one contour fringe corresponds to the magnetic flux of a vortex $h/2e$. On the right side of the picture, the magnetic line of a single vortex is seen protruding out of a superconducting lead surface and fanning out into space. This picture tells us that a magnetic vortex in lead is as small as 0.1 μm in diameter. To the left in the picture we can see an arc of a magnetic line. This is attributed to the vortex–antivortex pair predicted in the Kosterlitz–Thouless theory [16] in a two-dimensional system. The lines of magnetic flux from the two vortices are connected to form a circular line of magnetic flux.

The control of vortices in a superconductor is crucial in practical applications of superconductivity. For example, when a current is applied to a superconductor, a force is exerted on the vortices. If the vortices begin to move because of this force, heat is generated. The temperature of the superconductor then increases, and finally superconductivity breaks down. If we wish to keep superconductivity, we have to fix vortices against this force. How can we pin down the vortices? We can do it, for example, by introducing defects in the superconducting materials. However, it is only by trial and error that suitable pinning centres have been found. The mechanism of flux pinning is not yet well understood.

It would be very helpful for the study of pinning if we could see

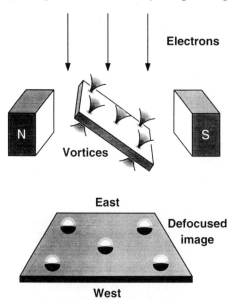

Fig. 10 Principle behind Lorentz microscopy. Incident electrons are deflected by the magnetic field of a vortex, and shifted in the lower plane thus producing a pair of bright and dark contrast regions.

directly the behaviour of the vortices in superconductors! To see the arrangement and the dynamics of vortices inside superconductors, we used Lorentz microscopy, another way of using the wave nature of electrons in electron microscopy. Lorentz microscopy is more suitable than holography for observing the dynamics of vortices. There were many unsuccessful attempts to observe vortices by Lorentz microscopy before the development of a coherent electron beam, because a highly collimated electron beam is indispensable to the technique. The principle behind Lorentz microscopy is rather simple (Fig. 10). Parallel electrons are incident on a vortex and are deflected slightly by the magnetic flux of the vortex. If the magnetic flux points to the south and the electrons fall from above, they are deflected westwards. When the electron intensity distribution is observed in the observation plane behind the specimen plane, more electrons arrive on the west side of the vortex and fewer electrons arrive on the east side. Therefore, the west side becomes brighter and the east side darker. We can tell the orientation of each vortex from the dark-to-bright orientation of its corresponding spot.

An example of vortex observation [17] is shown in Fig. 11. The sample is a niobium thin film cooled down to temperatures as low as −270 °C. The tiny spots are images of vortices. The dark lines are called

Fig. 11 Lorentz micrograph of vortices in niobium film at 4.5 K and 100 G. Tiny spots correspond to vortices. On careful observations a spot consists of dark and bright regions. The orientation of the vortex is determined by the line dividing the two regions. The dark lines are bend contours where incident electrons are scattered strongly owing to Bragg reflection.

Fig. 12 Lorentz micrograph of vortices in niobium film having surface steps. The vortices move along steps, but cannot cross them owing to pinning effects.

bend contours. They are caused by the bending of a thin film, which results in the strong scattering of electrons, called Bragg reflection, at the crystal lattices. Look at the spots carefully, and you will notice that each one has a bright side and a dark side.

When the applied magnetic field or the sample temperature is changed to a new value, the vortices begin to move until they settle in new stable positions. Using Lorentz microscopy, we could actually see this motion of vortices for the first time. They were found to move as if they were living creatures.

An example [18] of such vortex movement is shown in Fig. 12. The sample is a niobium thin film with surface steps where the film thickness changes. When the applied magnetic field changes, the vortices begin to move from the lower left to the upper right in the picture. Unable to climb up the step, they change their direction of motion and move along the step. The vortices move together in lines along the surface steps. When they reach a corner of the step, they turn right in two different directions. These movements show that surface steps act as pinning lines for vortices moving across them.

Conclusions

Electron waves, which play an important role in the microscopic world of atoms and molecules, can now be used to unveil the hitherto invisible microcosmos as a result of the development of a 'coherent' field-emission electron beam.

References

1. Zernike, F. (1992). *Physica*, **9**, 686.
2. Tonomura, A., Matsuda, T., Endo, J., Todokoro H., and Komoda, T. (1979). Development of a field emission electron microscope. *J. Electron Microsc.*, **28**, (1), 1.
3. Tonomura, A. (1993). Electron Holography, *Springer Series in Optical Sciences*, Vol. 70. Springer, Heidelberg.
4. Davisson, C. J. and Germer, L. H. (1927). *Phys. Rev.*, **30**, (6) 705.
5. Thomson, G. P. (1928). *Proc. R. Soc. London, Ser. A*, **117**, 600.
6. Kikuchi, S. (1928). *Proc. Imp. Acad.*, **4**, 354.
7. Feynman, R. P., Leighton, R. B., and Sands, M. (1963). *The Feynman Lectures on Physics*, Vol. III, p. 1–1. Addison Wiley, Menlo Park, CA.
8. Einstein, A., Podolsky, B., and Rosen, N. (1935). *Phys. Rev.*, **47**, 777.
9. Bohm, D. (1952). *Phys. Rev.*, **85**, 166.
10. Aspect, A. (1991). *Europhys. News*, **22**, 73.
11. Tonomura, A., Endo, J., Matsuda, T., Kawasaki, T., and Ezawa, H. (1989). *Am. J. Phys.*, **57** (2) 117.
12. Möllenstedt, G. and Düker, H. (1956). *Z. Phys.*, **145**, 375.
13. Gabor, D. (1949). *Proc. R. Soc.*, **A197**, 454.
14. Tonomura, A., Matsuda, T., Endo, J., Arii, T., and Mihama, K. (1980). *Phys. Rev. Lett.*, **44** (21), 1430.
15. Matsuda, T., Hasegawa, S., Igarashi, M., Kobayashi, T., Naito, M., Kajiyama, H., Endo, J., Osakabe, N., Tonomura, A., and Aoki, R. (1989). *Phys. Rev. Lett.*, **62** (21), 2519.
16. Kosterlitz, J. M. and Thouless, P. (1973). *J. Phys. C*, **6**, 1181.
17. Harada, K., Matsuda, T., Bonevich, J., Igarashi, M., Kondo, S., Pozzi, G., Kawabe, U., and Tonomura, A. (1992). *Nature*, **360**, 51.
18. Harada, K., Kasai, H., Matsuda, T., Yamasaki, M., and Tonomura, A. (1994). *Jpn. J. Appl. Phys.*, **33**, 2534.

AKIRA TONOMURA

Born 1942, educated at the University of Tokyo. After completing a B.S. in 1965, he joined the electron-beam physics group at the Hitachi Central Research Laboratory. In 1967, he embarked on a project to realize Professor Gabor's idea of electron holography. He succeeded in 1979 in developing a 'coherent' source of electron waves, an electron equivalent of laser in holography. Thus, the holography electron microscope materialized and has been constantly improved. It has unveiled many key phenomena in the physics of magnetism and superconductivity, and has put a stop to long-standing disputes over the reality of the gauge potential as embodied in the Aharonov–Bohm effect. He received a Dr.Eng. from Nagoya University in 1975 and a Ph.D. from Gakushuin University in 1993. He was awarded the Imperial Prize in 1991.

C_{60}: Buckminsterfullerene, the celestial sphere that fell to earth

HAROLD KROTO

Introduction

In the early 1970s, the chemistry of carbon in unsaturated configurations became the prime research focus for my group at the University of Sussex [1]. Free unstable species and reaction intermediates were made in which the oxygen atom of the carbonyl group was replaced by sulphur or selenium (i.e. $>C=S$ and $>C=Se$) and the nitrogen atom in imines and nitriles was replaced by phosphorus (i.e. $>C=P-$ and $C\equiv P$). Essentially, the objective was the creation of new compounds in which carbon was multiply bonded to sulphur, phosphorus or silicon and in some cases to another carbon atom. Until this time the so-called double bond rule suggested that double bonds to second row elements were unlikely. By thermolysis we found that by working at low pressure it was relatively easy to make and study the properties of compounds with $>C=S$, $>C=Se$, $>C=P-$, $-C\equiv P$, $-B=S$ and $-B=Se$ groups produced from specially synthesized precursors, and characterized by microwave and photoelectron spectroscopy [1]. We made the first carbon–phosphorus double bonded molecule $CH_2=PH$ and numerous derivatives as well as $CH_3C\equiv P$ the first derivative of $HC\equiv P$ and many other analogues [1–3]. These phosphaalkenes and phosphaalkynes were subsequently exploited further by several groups, including ourselves, as synthons.

In parallel with these studies we probed molecules containing carbon multiply bonded to itself in long polyyne chains. An earlier interest in carbon was reawakened in the early 1970s by work at Sussex on polyynes by my close friend and colleague David Walton. David had developed elegant methods for synthesizing long chain polyynes, based on the silyl protection techniques which he pioneered in acetylene chemistry [4–6]. He and his students had made the parent 24 carbon species

$$H–C\equiv C-C\equiv C-C\equiv C-C\equiv C-C\equiv C-C\equiv C-C\equiv C-C\equiv C-C\equiv C-C\equiv C-C\equiv C-C\equiv C-H$$

in minute quantities in solution and even a 32 carbon atom (silyl-protected) polyyne [5]. These chains were precisely what was needed for a study of vibration-rotation dynamics—a topic which in the early 1970s attracted my interest. In my mind such chain molecules conjured up an image of a microscopic cheerleader tossing a very bendy quantum-mechanical bamboo baton high into the air—and then attempting to catch it as it descended, flexing violently, and rotating at the same time. The obvious first compound for microwave study was the cyanopolyyne, HC_5N, which would possess a very large dipole moment. Walton devised a synthetic route and Andrew Alexander (Sussex Chemistry by Thesis student) tackled the very demanding problem and successfully prepared the molecule and measured its microwave spectrum in 1974 [7].

At about this time, spectacular advances were being made in molecular radioastronomy. The black clouds which are smeared across our Milky Way Galaxy (Fig. 1) were found to possess long-hidden dark secrets. Townes and co-workers [8] opened the Pandora's box in 1968 and revealed that the clouds were full of identifiable molecules. Thus space was no longer a playground preserved for astronomers, it now presented chemists with a novel piece of apparatus, indeed a colossal new spectroscopic sample cell containing a plethora of exotic molecules in a wide range of physicochemical environments [9]. I wrote to my friend and former colleague Takeshi Oka at the NRC with a view to collaborating in a radio search for HC_5N in space and in November 1975 the search (with Canadian astronomers Lorne Avery, Norm Broten and John McLeod) resulted in the successful detection of a signal from $SgrB_2$, a giant molecular cloud near the galactic centre [10]. This was a surprise as, in 1975, molecules with more than three or four heavy (C, N or O) atoms were assumed to be far too rare to be detectable. However, having found HC_5N in space, perhaps HC_7N was also present, so Walton devised a synthesis of HC_7N and graduate student Colin Kirby managed the very difficult problem of making it and measuring its spectrum [11]. We were actually already working on the radio telescope as our allotted observing session had started by the time Kirby (in England) finally succeeded in obtaining the vital frequency in the laboratory. He telephoned my wife and she then telephoned Fokke Creutzberg, a friend in Ottawa who, having also noted it down carefully, transmitted it to us at the telescope site in Algonquin Park (Fig. 2). The next few hours were high drama. We dashed out to the telescope and tuned the receiver to the predicted

Fig. 1 The dark clouds in Taurus from Barnards 1927 Survey *Atlas of selected regions of the Milky Way* (ed. E. B. Frost and R. Calvert, Carnegie Institute of Washington 1927). Heiles Cloud 2 is near the left hand bottom corner.

frequency range as Taurus (Fig. 1) rose above the horizon (perfect timing). We then tracked the ultra-weak signals from the cold dark cloud throughout the evening. We watched the oscilloscope for the slightest trace of the predicted signal in the receiver's central channel. By 01.00 hours the next morning we were too excited and impatient to wait any longer and, shortly before the cloud finally vanished completely, Avery stopped the run and processed the data. The moment, when the trace in Fig. 3 appeared on the display, was one of those that scientists dream about and which, at a stroke, compensate for the years of struggle and the plethora of disappointments endemic to a scientific life. The revelation that we had detected such a large molecule in space [12] was most exciting. The next candidate was HC_9N, but its synthesis was a daunting task. Fortunately Oka developed a beautifully simple extrapolation technique which predicted the possible radio frequencies of HC_9N and, almost unbelievably, we detected this chain as well [13]. It is not so obvious today but, in 1975–8, such long chains were a totally new and unex-

Fig. 2 The 46 m radiotelescope in Algonquin Park, Ontario, Canada with which the long carbon chain molecules in space were detected.

pected component of the interstellar medium and provided a major puzzle in that it was not at all clear how such species came to be present. The source of these long molecules became something of a preoccupation (perhaps even an obsession) and, by the early 1980s, I became convinced that the charismatically named red giant carbon stars must hold the key. Particularly interesting was the spectacular infrared object IRC + 10216 which pumps vast quantities of chains and grains out into space. Perhaps some symbiotic chain/dust carbon chemistry was involved [1, 9].

The foregoing account sets out some of my ideas about carbon in space at the time of a visit with Bob Curl at Rice University, Texas during Easter 1984. During this visit I was introduced to the remarkable

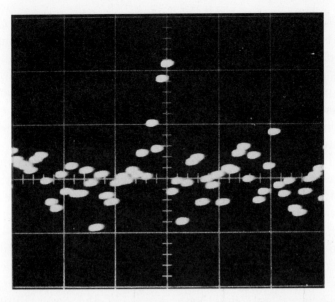

Fig. 3 A photograph of the first output of the radio signal due to interstellar HC_7N in Heiles's Cloud 2 in the constellation of Taurus, in Fig. 1.

laser vaporization cluster beam apparatus which Rick Smalley and co-workers had recently developed [14]. This powerful technique enabled refractory clusters, formed in a helium entrained plasma, produced by a pulsed laser focused on a solid target, to be studied by mass spectrometry. The technique was a major breakthrough in cluster science as it made refractory clusters accessible for detailed study for the first time. To see the apparatus in operation was quite fascinating and during the visit the thought crossed my mind that by using a graphite target it might be possible to simulate the chemistry which takes place in the atmosphere of a carbon star. Perhaps we could produce chains with as many as 24–32 carbon atoms, related to Walton's polyynes, or the equally interesting pure carbon species, such as C_{33}, seen by Hintenberger and his colleagues in the early 1960s, in a mass spectrometric study of a carbon arc [15]. There was also the prospect that we should be able to check out the fascinating idea of the late Alec Douglas [16], that carbon chains might be carriers of the diffuse interstellar bands—a set of absorption features [17] which has puzzled astronomers and spectroscopists for more than six decades.

The project was not considered a priority by the Rice cluster group at that stage, and it had to await a convenient slot in their programme. In the interim, an Exxon group carried out the basic graphite vaporization experiment and, in the summer of that year (1984), published the fascinating result that a totally new family of carbon clusters, C_{30}–C_{190}, formed [18]. This exciting discovery was peculiar because only even numbered clusters were detected. It is important to note that, at this juncture, no specific cluster was perceived to be special [18]. In late August 1985 (seventeen months after my visit to Rice), Curl telephoned to say that, at long last, my carbon chain study was scheduled to start almost immediately. As I was keen to direct my experiment personally, I packed my bags and arrived at Rice as soon as I could.

Silicon and germanium cluster studies—because of their semiconductor implications—were on the Rice plate when I arrived. The carbon project was to be completed quickly so as to cause as little delay as possible to the semiconductor (applied!) cluster programme. Preliminary measurements on carbon had however already been carried out—nothing untoward was noticed although, in retrospect, there were differences from the results published in the original Exxon study. My experiments started on Sunday, 1 September 1985. I worked in the laboratory alongside research students Jim Heath, Yuan Liu and Sean O'Brien. The experiments almost immediately confirmed the conjecture [1, 9] that long chain species can form in a carbon plasma similar to that in red giant stars. However as the experiments progressed it gradually became clear that something quite astonishing was taking place; as we varied the conditions from one run to the next, we noticed that a 720 amu peak (corresponding to a carbon cluster with sixty atoms) behaved in a most peculiar fashion. Sometimes it was completely off-scale, whereas at other times it was quite unassuming. The mass spectrum recorded on Wednesday, 4 September 1985 was astounding and I annotated our reactions on my copy of the printout (Fig. 4) and Heath, Liu and O'Brien also noted them in the laboratory record book (Fig. 5). On Friday, 6 September O'Brien found conditions under which C_{60} was 30 times stronger than the adjacent C_{58} (Fig. 5(b)). During Saturday and Sunday, Heath optimized conditions even further until he produced the final spectrum (Fig. 6) in which C_{60} is nearly 40 times more abundant than C_{62}. C_{70} is also prominent!

What on earth could C_{60}(?) be? As the days passed the quest for a structure which could account for this observation engendered constant discussion which became particularly intense during Monday, 9 September by which time the consensus was that C_{60} might be some sort of spheroidal cage. One possibility considered was that flakes of hexagonal carbon had been blasted from the graphite surface by the laser and that these hot graphene networks had somehow managed to wrap themselves up into closed cages, thus eliminating the dangling edge bonds and making the cage unreactive. This idea had instant appeal for

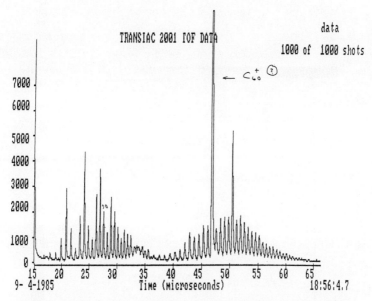

Fig. 4 Annotated time-of-flight mass spectrum of carbon clusters produced on Wednesday, 4 September 1985 the day on which the dominance of the C_{60} signal was first recorded (see Fig. 5(a)).

all. For me this concept brought back vivid memories of Buckminster Fuller's geodesic dome at EXPO 67 in Montreal (Plate 1). I had actually been inside this incredible structure 18 years before and remembered pushing my small son, in his pram, along the ramps and up the escalators, high up among the exhibition stands and close to the delicate *hexagonal* network of struts from which the edifice appeared primarily to have been constructed. This experience had left an image which could never be erased from my mind. I had collected numerous photographs of the dome from magazines over the years and particularly striking were those in my favourite graphic art and design magazine *Graphis*. As I remembered it from the *Graphis* pictures (Plate 1) the dome consisted of a plethora of hexagonally interconnected struts. The thought struck me that, as far as C_{60} was concerned, Buckminster Fuller's domes might provide some clue, and when I mentioned this to Smalley, who also favoured some sort of a 'chicken-wire' cage, he thought it would be a good idea to get a book on Fuller from Rice library. The second image that came to mind, most vividly on the Monday, was of a polyhedral cardboard stardome, Plate 2. I had constructed this map of the stars in the sky many years before when my children were young, but now it was tucked away in that universal quantum of storage — a Xerox box—somewhere downstairs in my home far away in England. I remembered

a)

b)

c)

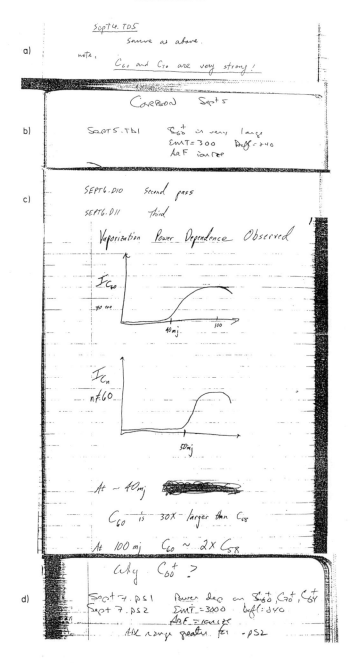

d)

cutting out not only hexagons but also pentagons and I described the main features to Curl as we walked home for lunch. I itched to get my hands on it and wondered whether I should telephone my wife to ask her to count the vertices in order to ascertain whether, as I half-suspected, they totalled sixty. I also had a third idea involving four

Fig. 5 Entries by Heath, Liu, and O'Brien in the Rice Cluster Laboratory Notebook from the period 4–7 September 1985 when key experiments were carried out. (a) September 4th; first recorded note that C_{60} and C_{70} were very strong (see also Fig. 4). (b) September 5th: the C_{60} signal is again noted as very strong. (c) Sept 6th; the record of first experiment which specifically aimed at the optimization of the conditions for the production of a dominant C_{60} signal – 'C_{60} is 30 × larger than C_{58}'. (d) September 7th; Notes on the careful and extensive studies which and provided the conclusive evidence (Fig. 6) of the special nature of C_{60}.

graphene layers involving 6/24/24/6 atoms which even came up with the number 60 [19].

I was due to return to the UK the next day so that evening (Monday, 9 September) I thought we should celebrate our extraordinary discovery and invited the group out to dinner at what had become our favourite Mexican restaurant. During the meal, the conversation was naturally dominated by C_{60}(?). We again considered all possible structures that had been thrown up during our deliberations over the past few days. We converged unanimously on a *closed cage*. We talked about Buckminster Fuller's domes, Smalley talked about chicken-wire cages, I reiterated the essentials of the stardome—its spheroidal shape and the fact that in addition to hexagonal faces it had pentagonal ones as well. After the

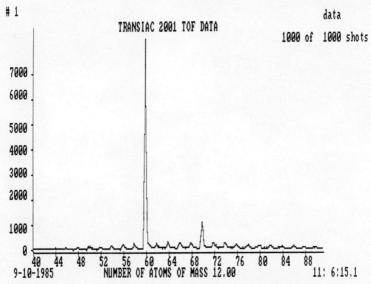

Fig. 6 Time-of-flight mass spectrum of carbon clusters under the optimum conditions for the observation of a dominant C_{60} (and C_{70}) signal.

meal, the other members of the group went home; I returned to the laboratory in order to study the Marks' book on Buckminster Fuller, but I could not find it. I again considered telephoning home about the stardome vertices, but it was now very late as it was the early hours of the morning in the UK.

Very early the next morning Curl telephoned to say that Smalley had experimented with geodesic paper models based on the hexagonal and *pentagonal* stardome characteristics I had described the previous day and that I must come to the lab as soon as possible. When I got to the laboratory and saw the paper model (Plate 2(b)) which Smalley had constructed during the night I was ecstatic. It seemed identical with the stardome, as I remembered it, and I was overtaken by its beauty, and particularly pleased that two of my three conjectures had been right all along. Heath and his wife Carmen had also experimented with a geodesic structure, modelled out of jelly beans and toothpicks; it was not quite so convincing [20]. The structure elucidation had been, just as the discovery experiment itself, a most satisfying synergistic team effort.

Then we found out that the C_{60} structure we were proposing had the same symmetry as the modern soccer ball (US) or football (Rest of the World). We immediately purchased a football and—appropriately—the five-a-side team was photographed (Fig. 7). At the moment that we came

Fig. 7 The five-a-side Rice/Sussex soccer/football team: O'Brien, Smalley, Curl, Kroto, and Heath.

Plate 1 Photograph taken by Robin Whyman of the Geodesic Dome designed by Buckminster Fuller for the US exhibit at Montreal EXPO67. One of the pentagons necessary for closure is discernible in this beautiful photograph.

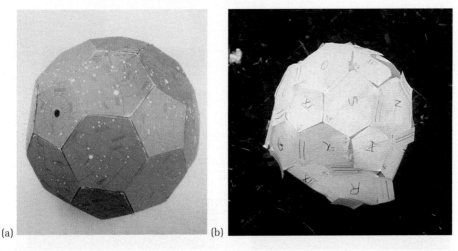

(a)

(b)

Plate 2 The two card models which played key roles in the positing of the truncated icosahedral structure of C_{60} (see text). (a) Stardome map of the sky (Buckminster Fuller patented t-icosahedral, and other polyhedral, world map projections); (b) the prototype model constructed by Smalley. Both models are truncated icosahedra with 60 vertices, 12 pentagonal and 20 hexagonal faces.

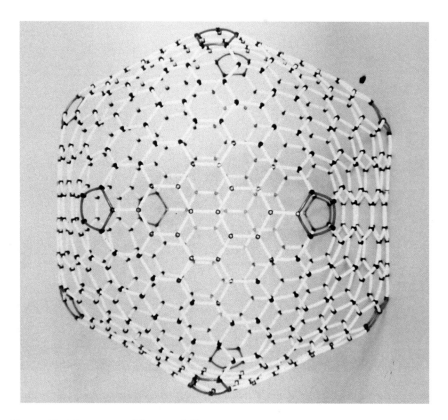

Plate 3 Model structure of the giant fullerene C_{240}.

Plate 4 Original 'red(d)ish' extract obtained on Monday, 6 September 1990.

Plate 5 Chromatographic separation by Taylor produced the magenta solution of C_{60} (left) and red solution of C_{70} (right).

to type the title of the discovery paper [21] into the PC, I plucked the name *Buckminsterfullerene* out of air thick with the *Monty Python* scripts we had plagiarized for decades to write departmental cabarets at Sussex. It seemed a highly appropriate name as the geodesic dome concept had played such an important part in arriving at the structural solution—at least as far as I was concerned. It was also fortuitous and rather satisfying that the -ene ending allowed the name to end nicely with the correct scientific terminology and furthermore flow smoothly off the (English!) tongue. At the time Smalley and Curl were not at all keen on my somewhat idiosyncratic proposal and took some convincing; fortunately, in the end, they agreed. As time has passed, it is pleasing to note how many lay people have been become interested in the molecule in particular and its chemistry in general, as a consequence of this some-what controversial name. It has also turned out to be most appropriate in that the whole family of closed cages can be appropriately named—Fullerenes [22].

The quest for macroscopic amounts of C_{60}

The 10 hour non-stop flight home to England was a physical and psycho-logical high for me. My first action on entering the house was to open the Xerox box and take out the stardome which had influenced our deliberations during the search for the structure of C_{60} [23]. It had always looked beautiful; now it positively glowed (Plate 2a). As the news of our sensational discovery spread, Martyn Poliakoff at Nottingham wrote to say that his friend, David Jones, had already thought of the possibility of hollow carbon cages. Jones had written a delightful article in the *New Scientist* (1966) under the pseudonym Daedalus [24, 25] and had suggested that the high temperature graphite production process might be modified to generate graphite balloons. I thus learned of the consequences of Euler's law—that 12 pentagons were needed to close a network, hexagons alone just will not do. I was introduced to D'Arcy Thompson's elegant book [26] and realized that one could close up an even-numbered carbon cage with any number of hexagons (except one, I subsequently learned) provided 12 pentagons were included. We also learned that Osawa and Yoshida (1970) had already thought of C_{60} and had even suggested that it might be superaromatic [27, 28]. Bochvar and Gal'pern [29] had also published a paper on the molecule two years later. Indeed, Orville Chapman at UCLA had even initiated a programme to synthesize the molecule [30]. Thus it transpired that although some most imaginative scientists were around, almost no-one had noticed their pioneering flights of imagination.

The first experimental question stimulated by the realization that we might have found a 'third' form of carbon with a hollow, cage structure was obvious: is it possible to trap an atom inside the cage? Heath made the key breakthrough when he succeeded in detecting a stable $C_{60}La$ complex [31]. This result provided the first concrete piece of evidence to support our cage proposal.

My attitude during the late 1970s and early 1980s had been that it was simply not enough just to detect the carbon chains; I continually wondered about possible ways in which the circumstellar route to the chains might be confirmed experimentally and it was this recurring obsession which initiated the experiments which uncovered C_{60}. After 1985, I developed an analogous attitude to C_{60}. The discovery that it *probably* was a truncated icosahedron was 'not enough' and as time passed, confirmation of the structure and its identification in space also became somewhat of an obsession. We (the Rice group and I) were convinced from the outset that our proposal was correct—it was just too beautiful to be wrong—however if, for once, aesthetics were misleading us, it seemed to me that it would be much better if we proved it right or wrong ourselves. It is certainly true that though many loved C_{60} and were convinced that we were right, some greeted our proposal with scepticism. During this period (1985–90) I examined the problem independently at Sussex, as well as in collaboration with the Rice group, who also made a series of independent studies. All these efforts in time resulted in a mass of convincing circumstantial evidence in support of our proposal [32, 33].

From the moment of our discovery that C_{60} was stable, I had really only one specific dream—that of solving the molecule's structure by NMR spectroscopy. As all 60 carbon positions in Buckminsterfullerene are equivalent, the ^{13}C NMR signal should consist of a single line. Proof of our radical proposal in such an exquisitely simple manner remained dormant for some five years; almost like a holy grail—apparently far, far beyond reach. As time passed, several advances were made [32, 33], and the cage concept often seemed to clarify previously known observations. That was for me an important criterion. I remembered hearing Richard Feynman on BBC radio saying that it seemed to him that if a new radical theory were right it would allow previously known puzzles to fall into place. I decided to quantify this: if an idea fitted 80–90 per cent of the puzzling observations, it was probably right; if it had to be bent to fit more than 10–20 per cent it was almost certainly wrong. Buckminsterfullerene was well up in the first category.

I remember vividly the day on which all my doubts vanished. I was sitting at my desk thinking about the reasons why C_{60} might be stable.

The final piece of evidence in my own personal jigsaw puzzle turned out to be so simple that it was almost child's play, indeed it evolved by playing with molecular models of various conceivable cages. The solution eventually revolved around the C_{70} signal which always popped up prominently when C_{60} was strong. I used to call them the Lone Ranger and Tonto because they were so inseparable and C_{60} was always the stronger and C_{70} always the weaker (Figs 4 and 6). As far as C_{60} was concerned, it seemed to me that it could be no accident that the modern soccerball had the same layout. Maybe the football held a simple clue. One of the most striking and fascinating points, at least visually, when I looked carefully at a football, was the fact that all the (black) pentagons are isolated, whereas all the (white) hexagons are linked. Now it is well known that compounds with abutting pentagons (the pentalene configuration), without substituents to facilitate extended conjugation, tend to be unstable. Furthermore once I had learned (from Daedalus and Thompson [24–26]) that no matter how many hexagons there were, Euler's law showed that 12 pentagons were necessary for closure, it was also obvious that C_{60} must be the smallest closed carbon cage with all pentagons isolated—because $5 \times 12 = 60$. As far as C_{60} was concerned, for me, this seemed to be the equivalent $2 + 2 = 4$. The reactivity experiments which O'Brien mainly had carried out [31] had convinced us that closure was possible, even probable, for all the even clusters. I realized that a *pentagon isolation* could account for the stability of C_{60} in this simple and elegant manner. I then began to wonder which cage might be the next for which pentagon isolation was feasible. Playing with the models indicated immediately that it was not C_{62} and, as I added more atoms, try as I would I could not find cage structures for C_{64}, C_{66} and C_{68} which had isolated pentagons. We had already proposed a structure for C_{70} [31] after Smalley had shown that by splitting C_{60} into two C_{30} hemispheres a ring of ten extra carbons could be neatly inserted, producing a most elegant symmetric egg-shaped structure, Fig. 8(b). It then began cross my mind that perhaps this C_{70} cage might be the *second* isolated pentagon cage. This surprised me and then it dawned on me that if this were correct it would explain the fact that C_{70} was also prominent (Figs 4 and 6) beautifully. I then became excited as I realized that if the conjecture really were correct it would provide the most elegant and compelling support yet available for the whole concept of closure in general and *the Buckminsterfullerene structure of C_{60} in particular*. We had stuck our necks out with C_{60} on the basis of a single mass spectrometric peak, but it now seemed that the cage proposal positively demanded that both C_{60} *and* C_{70} be special [34]. That would be wonderful. Another solution that demanded these *two* particular magic numbers seemed

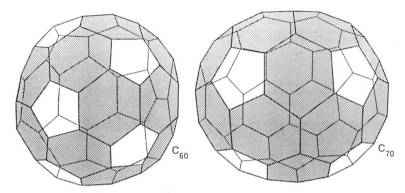

Fig. 8 For species with less than 72 atoms these two structures are the *only* ones which can be constructed without abutting pentagons. Thus on the basis of the *pentagon isolation principle* and geodesic considerations C_{60} and C_{70} are predicted to be the first and second fullerene magic numbers. This result together with Fig. 6 provided the simplest and strongest circumstantial evidence supporting the closed cage concept.

impossible, especially as 70 would be a very peculiar number to be deduced as magic. I knew that Nature would not be that perverse and for the first time was absolutely convinced our structure was correct and never again doubted that one day we would be vindicated. Indeed as far as I was concerned this result meant that we were home and dry. I contacted Schmalz, Klein, Hite and colleagues in Galveston to ask them whether they could help me to check out whether my conjecture was right, only to discover that they had already come to a similar conclusion from their quantitative studies [35].

One Sunday afternoon, with molecular models strewn all over the coffee table at home in Sussex, I decided to investigate other possible fullerene structures, in particular the structure of an interesting C_{32} cage which the results of O'Brien *et al.* [31] had suggested might be the smallest cage that could exist. As I counted up the carbon atoms on the structure I had guessed might explain this result, I realized that it had 28 atoms—four too few. In fact it was a delightfully symmetric C_{28} cage (Fig. 9) and suddenly I wondered whether it might explain a highly dominant C_{28} signal which we had occasionally noticed. Further model building uncovered a possible explanation of the fact that under certain conditions the set C_{24}, C_{28}, C_{32}, C_{50}, C_{60} and C_{70} were all magic. C_{28} (Fig. 9) now became my own personal favourite because, along whichever of the four threefold axes one views this tetrahedral molecule, it looks uncannily like Gomberg's famous free radical, triphenylmethyl; this also just had to be right. It had always amused me to think that on

Fig. 9 Photograph of the molecular model of C_{28}—one of several fullerenes (with 24, 28, 32, 50, 70 atoms) predicted to be stable on the basis of geodesic and chemical considerations[35]. These magic number predictions fitted almost perfectly those observed under certain clustering conditions[43].

attempting to make the rather mundane compound hexaphenylethane, Gomberg failed (miserably), produced the triphenylmethyl radical and had to console himself with becoming known as the Father of Free Radical Chemistry instead. This sort of failure appealed to me. Furthermore the cage would have lone sp^3-type electrons at the four tetrahedral vertices and that implied that $C_{28}H_4$ should be a fairly stable hydrofullerene—almost a tetrahedral cluster analogue of the sp^3 carbon atom itself, a sort of carbon supercluster atom.

At this time I had a telephone conversation about nomenclature with Alex Nickon in which we came to the conclusion that the name 'fullerenes' would work well for the whole carbon cage family [22]. I was delighted with this refinement and especially with the fact that it was even more appropriate now as Buckminster Fuller had patented constructs of all shapes and sizes based on the 5/6 ring principle [36], some very similar to the elongated C_{70}.

One day I decided to buy £300 worth (ten thousand) of sp^2 carbon atoms (molecular models!) just because I wanted to build a giant fullerene such as C_{6000} (10 times the diameter of C_{60}) and see what it looked like. Ken McKay (graduate student) obtained Coxeter's book [37], and Goldberg's paper [38] and set to work. When he came into my room with C_{540} (Plate 3) I was delighted because the model was truly beautiful, but also somewhat perplexed because it was not perfectly round like Buckminster Fuller's domes. It was in fact interestingly different, essentially a monosurface which swept between cusps in the vicinity of the pentagons; it had icosahedral symmetry, but it was not an icosahedron [39]. Then we realized that the shapes would explain [39] some beautiful polyhedral patterns in graphite microparticle electron microscope pictures published by Iijima [40]. These microparticles could be nicely explained as concentric shells of graphitic carbon where the shells were giant fullerenes. Few of our discoveries have given me more personal delight than these shapes, partly because they are such elegant objects, but also because the exercise was started purely for aesthetic reasons with no serious aim and yet it had yielded such an important result.

Until late 1989, the evidence in support of our structure proposal was, to my mind, very convincing but clearly a sample of C_{60} was needed— not just a whisper in a helium wind. One project, initiated with Ken McKay, probed (by electron microscopy) the films produced by a carbon arc under helium in an old bell jar evaporator. We found that, as the helium pressure increased, the film microstructure changed. The quadrupole mass spectrometer, with which I sought to monitor whether or not C_{60} was produced in this experiment, was the integral part of a modest research proposal to prosecute this project effectively which failed to attract support from any funding source. Then in September 1989, Michael Jura (Astronomy Department UCLA), sent me a copy of a thought-provoking paper [41] presented by Krätschmer, Fostiropoulos and Huffman at a Symposium on Interstellar Dust. They had observed four weak but distinct infrared absorptions in a film of arc-processed graphite, tantalizingly consistent with Buckminsterfullerene. From theoretical studies [42] it was well known that C_{60} should exhibit only four lines. What was more, the observed frequencies agreed quite well with theory. I was sceptical about their result and also rather chagrined, for had not McKay and I made soot in a bell jar, in exactly the same way some two to three years before when I initiated the project for which I was *still* trying to get support?

However I was intrigued and decided to resurrect the old and decrepit bell jar and with Jonathan Hare tried to reproduce the infrared features at a time when a third year undergraduate project was needed. It matters little whether or not such projects yield results; they should imbue the student with the flavour and excitement of genuine research, that is, the experience of working in the dark—not, as all-too-often happens, a frantic scramble for results at all costs! I had often initiated the most speculative of projects in this way and experience had shown, time and time again, that important and exciting studies could often start from such inauspicious beginnings. It seemed perfect for Amit Sarkar's third year project and he joined Hare on this wildly speculative project. Fairly soon Hare and Sarkar succeeded in obtaining the bands (Fig. 10). The obsolete apparatus promptly fell to pieces and after rebuilding it Hare carefully varied every experimental parameter possible and finally developed the expertise necessary to produce films which exhibited the tell-tale infrared features consistently. We really needed another way to monitor the samples, so we tried FAB sampled mass spectrometry. Ala'a Abdul-Sada helped us and finally we obtained a mass spectrum,

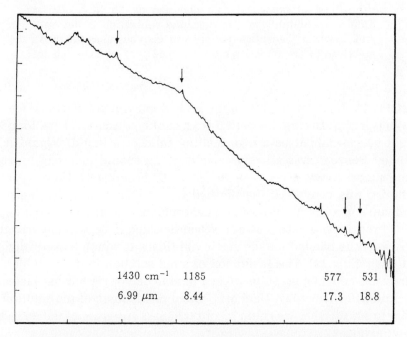

| 1430 cm^{-1} | 1185 | 577 | 531 |
| 6.99 μm | 8.44 | 17.3 | 18.8 |

Fig. 10 Infrared spectrum of a film of arc-processed carbon obtained by Hare and Sarkar at Sussex which shows weak but clear (and confirmatory) evidence that the features first observed by Krätschmer *et al.*[42] and tentatively assigned by them to C_{60} were repeatable.

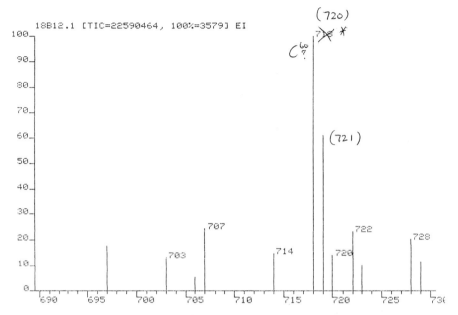

Fig. 11 Fast atom bombardment (FAB) mass spectrum of a deposit of arc-processed soot obtained on 23 July 1990 at Sussex by Abdul-Sada. The machine calibration is out by 2 amu however the isotope pattern was convincing as the peaks are close to the intensity ratio 1.0:0.66:0.22 as expected for $^{12}C_{60}$:$^{12}C_{59}^{13}C$:$^{12}C_{58}^{13}C_2$.

Figs 11, 12. Now for the first time we had convincing evidence for the presence of C_{60}. During this period we had often discussed what form C_{60} might take: would it be a high melting solid or a liquid? Would it be soluble? Benzene was a likely solvent as C_{60} should look like benzene from almost every aspect—or at least 20. Fortunately Hare, ever an optimist, was convinced that he had C_{60} and on Friday, 3 August he made up a mixture of the soot and benzene, in a small vial, which he set aside over the weekend (Fig. 12). When he came in on Monday morning (6 August) he handed me the small vial (Plate 4) which looked 'slightly red(d)ish' (Fig. 12). The generation of a red solution from insoluble and intractable common graphite seemed hard to believe. A few days later on the following Thursday, Hare evaporated down some of the solution in order to see whether we could record the mass spectrum of the extracted material (Fig. 12).

The next morning, Friday, 10 August, I received a telephone request from Philip Ball at the journal *Nature* to referee a paper by the Heidelberg/Tucson group. I was however totally unprepared for the BOMBSHELL which arrived at midday by fax. Wolfgang Krätschmer, Lowell

possible issue of FAB Mass Spect.   26/7/90.

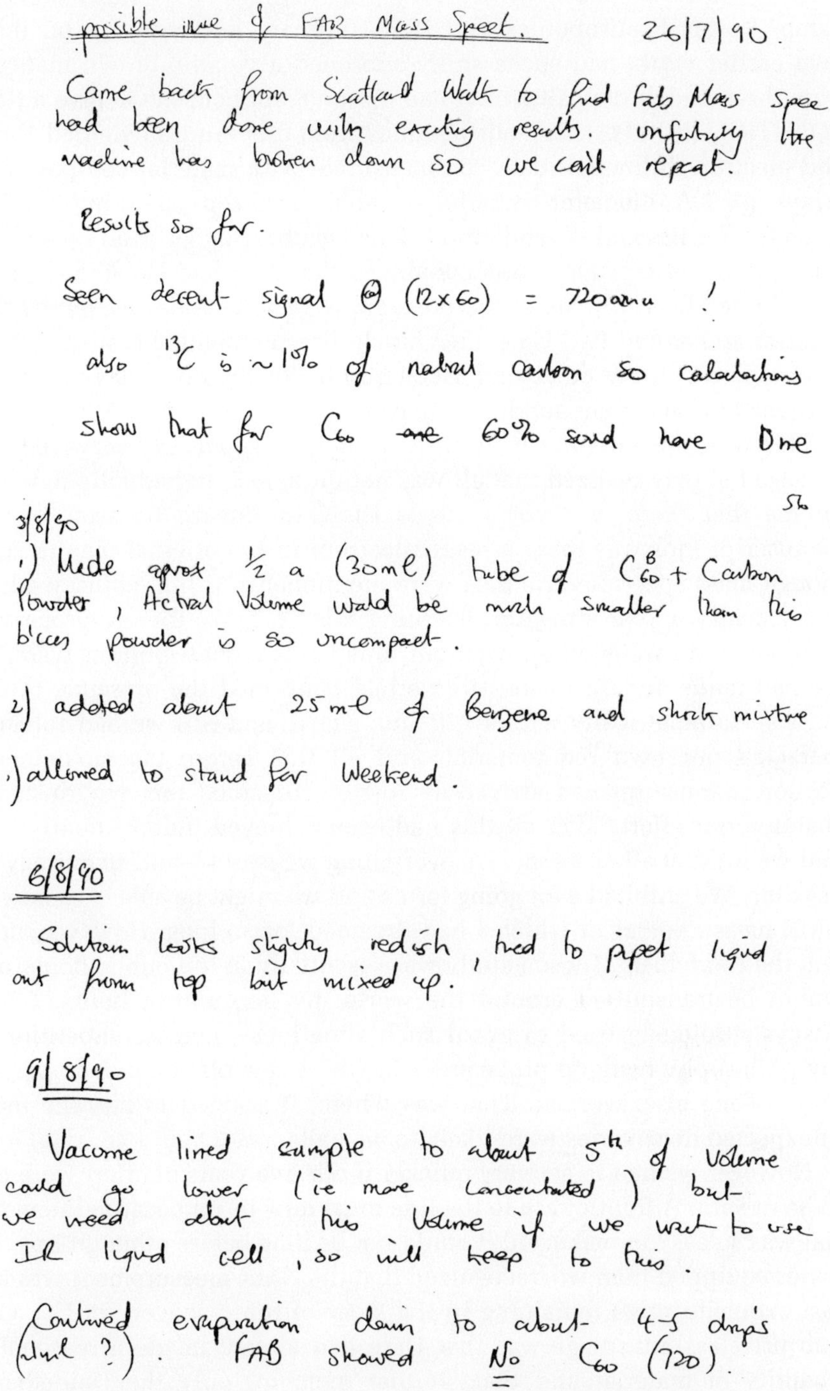

Came back from Scotland Walk to find fab Mass spec
had been done with exciting results. Unfortunatly the
machine has broken down so we can't repeat.

Results so far.

Seen decent signal @ (12×60) = 720 amu !

also ^{13}C is ~1% of natural carbon so calculations

show that for C_{60} one 60% sould have One

$\frac{3}{8}$90   5b

') Made aprox ½ a (30 ml) tube of C_{60}^B + Carbon
Powder, Actual Volume would be much smaller than this
b'cos powder is so uncompact.

2) added about 25 ml of Benzene and shook mixture

,) allowed to stand for Weekend.

6/8/90

Solution looks slightly redish, tried to pipet liquid
out from top but mixed up.

9|8|90

Vacume lined sample to about 5th of Volume
could go lower (ie more concentrated) but
we need about this Volume if we want to use
IR liquid cell, so will keep to this.

Continued evaporation down to about 4–5 drops
(dry ?). FAB showed No C_{60} (720).

Fig. 12 Entries by Hare in his laboratory notebook: (a) 26/7/90
(b) 3/8/90 (c) 6/8/90 (d) 9/8/90.

Lamb, Kostas Fostiropoulos and Don Huffman, following up on their own earlier work, had successfully sublimed a volatile brown material from their carbon deposit which had dissolved in benzene to give a RED SOLUTION (!!!!!) [43]. Crystals obtained from this solution yielded X-ray and electron diffraction data commensurate with material composed of arrays of 7 Å diameter spheroidal molecules, separated by 3 Å, as expected for Buckminsterfullerene. This beautiful paper even contained photographs of crystals. I was convinced that they had isolated C_{60} and that we had been pipped at the post. I was somewhat devastated by the disaster and called Ball back after lunch. I recommended that the paper be accepted without delay and asked him to convey my sincere congratulations to Krätschmer and his associates.

This was, needless to say, a difficult moment—but, as I surveyed the damage I slowly realized that all was not quite lost. It gradually dawned on me that there was not a single (!) NMR line to be seen in the manuscript (nor was there a mass spectrum in the original manuscript, though mass spectrometric data were mentioned). At this point, we had spent nearly a year struggling learning how to make the arc-processed material, and I weighed up what our independent contributions were: (i) we had made our own soot, (ii) we had confirmed the presence of C_{60} mass spectrometrically with a 720 amu signal, and (iii) we had solvent-extracted our own red material; and all this *before* the Heidelberg/Tucson manuscript had arrived at Sussex. I decided that we must not abandon our efforts after all this had been achieved independently and that we must at all costs salvage everything we could—and as quickly as possible. We still had a lot going for us and we might be able to make the NMR measurement of which I had dreamed for so long. However, now that the Heidelberg/Tucson study was essentially in the public domain it would be transmitted around the world, by fax, within hours. I had always studiously tried to avoid such situations—fierce competition—my philosophy being to probe areas in which few others, indeed preferably no-one else, worked. That was where, it seemed to me, the most unexpected discoveries were likely to be made.

However we had to act very quickly if our five years of effort were not to be in vain. A frantic race to the line must now be on because the material was so easy to make and it could not be long before other groups, far better equipped than we, recognized that the NMR measurement was the last exquisite prize remaining in the story of the discovery of C_{60}. Our one priceless advantage was that Hare had already made a reasonable quantity of material and that, at that moment, only the Heidelberg/Tucson group had any at all. We needed help and fast; Roger Taylor, a colleague at Sussex, provided the desperately needed expertise—Taylor separated the material quickly and efficiently and we obtained the mass spectrum shown in Fig. 13. As well as C_{60}, C_{70} was abundant.

Taylor found that the extract was soluble in hexane and realized that he might be able to separate the fullerenes chromatographically. To his delight he found that, on an alumina column, solution resolved into two bands—one red, the other magenta. The resulting magenta fraction (Plate 5), was a delight to the eye in the delicacy of its colour. It yielded a 720 mass signal and was sent to Tony Avent for NMR analysis. We were summoned to see our single line (Fig. 14) which Avent assured us was present. So this was it; a line so small a microscope was needed to see it! Could this insignificant little blip really be the line I had dreamed about for five years? Further work confirmed it beautifully as the result which we had sought, for so long in vain [44]. The joy alleviated almost all the despair I had first felt on reading the paper by Krätschmer *et al.* [41]. C_{70} turned out to be the icing on the cake, although not without attendant trauma. The wine red fraction, which had masked the pale magenta of C_{60}, yielded a mass spectral peak at 840 amu and was clearly C_{70} which should, according to the expected structure (Fig. 8), exhibit a five line ^{13}C NMR spectrum. Even more important was the fact that the spectrum confirmed the fullerene cages as a family concept in general.

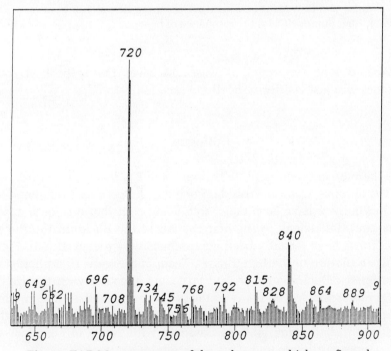

Fig. 13 FAB Mass spectrum of the red extract which confirmed that it consisted mainly of C_{60} and C_{70}.

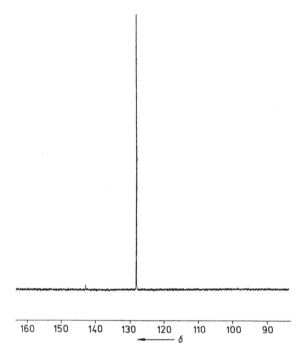

160 150 140 130 120 110 100 90
δ

Fig. 14 The first NMR trace in which the C_{60} resonance (at 143 ppm) was first identified (just!). The strong line at 128 ppm is (rather appropriately) the resonance of benzene.

As we had long suspected, it was now clear that a host of stable fullerenes was just waiting to be discovered [45].

Epilogue

Since the moment that graphite balloons and C_{60} itself were twinkles in the eyes of Jones, Osawa, Yoshida, Bochvar, Gal'pern and Chapman [24, 25, 27–30], there have been many significant contributions, both experimental and theoretical, to the first chapter of the Buckminsterfullerene Story. These have recently been comprehensively reviewed [45]. Figure 7 shows a picture of the Rice/Sussex Team and Figure 15 a photograph of the Sussex Buckaneers. These two teams, together with the Heidelberg/Tucson team scored many of the goals in the football match which has just ended. Many new teams have now started to play an exciting, but different, ball game [45]. At Sussex a wide range of chemical and materials properties are under investigation [46].

Apart from the success inherent in the observation of the 'lone' NMR

Fig. 15 Sussex Buckaneers Football Team. *From left*: Abdul-Sada and Hare (standing), Kroto, Taylor, and Walton (seated).

line, two matters afford a sense of deep pleasure. One was the fact that the colour of C_{60} was so very beautiful and, furthermore, that it was seen first at Sussex. The second was that we had extracted the material independently at Sussex. Indeed apparently no-one else, apart from us and Meijer and Bethune at IBM [47], appeared to have followed up the early Heidelberg/Tucson infrared observation (September 1989). In retrospect I find this astounding; perhaps it was because the work was published in the astronomy literature, but more probably it was because research, today, is carried out under 'applied' pressure and is imbued with the fear of failure. These factors are inimical to success in science for many scientists. The situation is exacerbated by the funding agencies which monotonously drone on so much about so-called 'wealth creation' that few researchers enjoy the luxury of really working in the dark. Only when you have no idea where the road leads does research embody the true spirit of scientific adventure and historically this approach has uncovered the true surprises and ultimately the great wealth-creating fundamental advances. A new Postbuckminsterfullerene World of Round Organic Chemistry and Materials Science has been discovered overnight. Almost every day, two or more papers appear on some novel aspect of fullerene behaviour. The origin of the whole programme arose from an interest in the fundamental aspects of molecular dynamics, allied to the quest for an understanding of the origin of the carbon chains in space and their possible relationship to circumstellar and interstellar grains as well as soot [1, 9]. Krätschmer and Huffman and their co-workers were originally motivated by a fascination with spacedust and, in their recent breakthrough, by the astrophysical implications of C_{60}. Finally it is worth noting that C_{60} should have been detected 30–50 years ago by applied scientists studying sooting flames! Thus the discovery of the fullernes is shining testament to the power of *pure fundamental* science and serves as a timely warning that it can achieve results where applied science has manifestly failed.

Acknowledgements

Important contributions were made in research programmes at the University of Sussex (Brighton, UK), the National Research Council (Ottawa, Canada), Rice University (Houston, Texas, USA), the Max Planck Institut für Kernphysik (Heidelberg, Germany) and the University of Arizona (Tucson, USA). The discovery of C_{60} is a tribute to not only the international nature of science but also the necessity of inter-disciplinary cooperation. The Sussex contribution to the story started as a consequence of the 'Chemistry by Thesis' degree course initiated by Colin Eaborn, which enabled undergraduates (such as Anthony Alexander) to carry out research with supervisors from more than one field. The Sussex contribution could not have been achieved had chemistry at Sussex been rigidly divided into those traditional subsections: organic, inorganic and physical. I also had advice from astronomers at Sussex—particularly Bill McCrea and Robert Smith. The carbon discoveries resulted from a research programme which started with synthetic chemistry (with David Walton, Anthony Alexander and Colin Kirby) and moved via spectroscopy and quantum mechanics to radioastronomy (with Takeshi Oka, Lorne Avery, Norm Broten and John McLeod at NRC). It moved back to earth and chemical physics (with Jim Heath, Sean O'Brien, Yuan Liu, Bob Curl and Rick Smalley at Rice). In the last phase, the key advice which came from Michael Jura (astronomer at UCLA) reactivated a project (with Jonathan Hare, Amit Sarkar, Ala'a Abdul Sada, Roger Taylor and David Walton) which I had originally started with Ken McKay. I also acknowledge greatly the help of many others, in particular Steve Wood of British Gas who helped us at the most critical time as well as many others, who played indirect parts, in particular Roger Suffolk, graduate students and post-doctoral fellows.

References

1. Kroto, H. W. (1982). *Chem. Soc. Rev.,* **11**, 435–91.
2. Hopkinson, M. J., Kroto, H. W., Nixon, J. F., and Simmons, N. P. C. (1976). *J. Chem. Soc. Chem. Commun.,* 513.
3. Hopkinson, M. J., Kroto, H. W., Nixon, J. F., and Simmons, N. P. C. (1976). *Chem. Phys. Lett.,* **42**, 460.
4. Eastmond, R. and Walton, D. R. M. (1968). *Chem. Commun.,* 204–5.
5. Eastmond, R., Johnson, T. R., and Walton, D. R. M. (1972). *Tetrahedron,* **28**, 4601–16.
6. Johnson, T. R. and Walton, D. R. M. (1972). *Tetrahedron,* **28**, 5221–36.
7. Alexander, A. J., Kroto, H. W., and Walton, D. R. M. (1976). *J. Mol. Spectrosc.,* **62**, 175–80.
8. Cheung, A. C., Rank, D. M., Townes, C. H., Thornton, D. D., and Welch, W. J. (1968). *Phys. Rev. Lett.,* **21**, 1701.
9. Kroto, H. W. (1981). *Int. Rev. Phys. Chem.,* **1**, 309–76.
10. Avery, L. W., Broten, N. W., Macleod, J. M., Oka, T., and Kroto, H. W. (1976). *Astrophys. J.,* **205**, L173–5.
11. Kirby, C., Kroto, H. W., and Walton, D. R. M. (1980). *J. Mol. Spectrosc.,* **83**, 261–5.
12. Kroto, H. W., Kirby, C., Walton, D. R. M., Avery, L. W., Broten, N. W., Macleod, J. M., and Oka, T. (1978). *Astrophys. J.,* **219**, L133–7.
13. Broten, N. W., Oka. T., Avery, L. W., Macleod, J. M., and Kroto, H. W. (1978). *Astrophys. J.,* **223**, L105–7.
14. Dietz, T. G., Duncan, M. A., Powers, D. E., and Smalley, R. E. (1981). *J. Chem. Phys.,* **74**, 6511–12.
15. Hintenberger, H., Franzen, J., and Schuy, K. D. (1963). *Z. Naturforsch.* Teil A, **18**, 1236–7.
16. Douglas, A. E. (1977). *Nature (London),* **269**, 130–2.
17. Herbig, G. H. (1975). *Astrophys. J.,* **196**, 129–60.
18. Rohlfing, E. A., Cox, D. M., and Kaldor, A. (1984). *J. Chem. Phys.,* **81**, 3322–30.
19. Ball, P. (1994). *Designing the Molecular World,* p. 43. Princeton University Press, Princeton, NJ.
20. Kroto, H. W. (1986). *Proc. R. Inst.,* **58**, 45–72.
21. Kroto, H. W., Heath, J. R., O'Brien, S. C., Curl, R. F., and Smalley, R. E. (1985). *Nature (London),* **318**, 162–3.
22. Nickon, A. and Silversmith, E. F. (1987). *Organic Chemistry—The Name Game: Modern Coined Terms and Their Origins.* Pergamon, New York.
23. Smalley, R. E. (1991). *The Sciences,* March–April, p. 22.
24. Jones, D. E. H. (1966). *New Sci.,* 3 November, p. 245.
25. Jones, D. E. H. (1982). *The Inventions of Daedalus,* pp. 118–19. Freeman, Oxford.
26. Thompson, D. W. (1942). *On Growth and Form.* Cambridge University Press.
27. Osawa, E. (1970). *Kagaku (Kyoto),* **25**, 854–63 (in Japanese); *Chem. Abstr.,* **1971**, **74**, 75698v.
28. Yoshida, Z. and Osawa, E. (1971). *Aromaticity* (in Japanese). Kagakudojin, Kyoto.

29. Bochvar, D. A. and Gal'pern, E. G. (1973). *Dokl. Akad. Nauk SSSR*, **209**, 610–12; English translation *Proc. Acad. Sci. USSR*, **1973, 209**, 239–41.
30. Chapman, O. (1985) Private communication.
31. O'Brien, S. C., Heath, J. R., Zhang, Q., Liu, Y., Curl, R. F., Kroto, H. W., and Smalley, R. E. (1985). *J. Am. Chem. Soc.*, **107**, 7779–80.
32. Kroto, H. W. (1988). *Science*, **242**, 1139–45.
33. Curl, R. F. and Smalley, R. E. (1988). *Science*, **242**, 1017–22.
34. Kroto, H. W. (1987). *Nature (London)*, **329**, 529–31.
35. Schmalz, T. G., Seitz, W. A., Klein, D. J., and Hite, G. E. (1988). *J. Am. Chem. Soc.*, **110**, 1113–27.
36. Fuller, R. B. (1983). *Inventions—The Patented Works of Buckminster Fuller.* St. Martin's Press, New York.
37. Coxeter, H. S. M. (1963). *Regular Polytopes.* Macmillan, New York.
38. Goldberg, M. (1937). *Tohoku Math. J.*, **43**, 104.
39. Kroto, H. W. and McKay, K. G. (1988). *Nature (London)*, **331**, 328–31.
40. Iijima, S. (1980). *J. Cryst. Growth*, **5**, 675–83.
41. Krätschmer, W., Fostiropoulos, K., and Huffman, D. R. (1990). *Dusty Objects in the Universe* (ed. E. Bussoletti and A. A. Vittone), pp. 89–93. Kluwer, Dordrecht.
42. Kroto, H. W., Allaf, W., and Balm, S. P. (1991). *Chem. Rev.*, **91**, 1213–35.
43. Krätschmer, W., Lamb, L. D., Fostiropoulos, K., and Huffman, D. R. (1990). *Nature (London)*, **347**, 354–8.
44. Taylor, R., Hare, J. P., Abdul-Sada, A. K., and Kroto, H. W. (1990). *J. Chem. Soc. Chem. Commun.*, 1423–5.
45. Taylor, R. and Walton, D. R. M. (1993). *Nature*, **363**, 685–93.
46. Birkett, P. R., Crane, J. D., Hitchcock, P. B., Kroto, H. W., Meidine, M. F., Taylor, R., and Walton, D. R. M. (1993) *J. Mol. Struct.* **292**, 1–8.
47. Meijer, G. and Bethune, D. S. (1990). *Chem. Phys. Lett.*, **175**, 1–2.

HAROLD KROTO

Born 1939, educated at Bolton School. After obtaining a B.Sc. and Ph.D. at the University of Sheffield, he went to the National Research Council, Ottawa in 1964 to continue spectroscopic studies of free radicals by flash photolysis and apply microwave techniques to unstable molecules. In 1966, after a year at the Bell Telephone Laboratories in New Jersey, carrying out Laser Raman and Quantum Chemistry research, he went to the University of Sussex, where he has developed spectroscopy and radio-astronomy techniques to study unstable species of importance in chemistry and astrophysics. His present work focuses on the chemistry and materials science of fullerenes and fullerene-related structures. He became a Lecturer in 1968, Reader in 1977 and Professor in 1985. He is now a Royal Society Research Professor. Prof. Kroto was knighted for his contribution to chemistry in 1996.

How much cosmology should you believe?

MARTIN REES

The big bang, dark matter, ripples in space and quasars get their full share of media attention.

But can you believe all the cosmology reported in the newspapers? In this paper, I describe what some recent observations tell us about how our Universe has evolved, and then venture into more speculative territory, where we have conjectures rather than consensus. I try throughout to distinguish between what is quite well established and what is not.

The Universe of galaxies

A word first about the galaxies. Our Milky Way, with its hundred billion stars, is just one galaxy similar to millions of others visible with large telescopes. Andromeda—a spinning disc viewed obliquely—is a nearby galaxy very like our own.

Galaxies are held in equilibrium by a balance between two effects—gravity, which tends to make the stars all gather together, and the countervailing effect of the stellar motions, which if gravity did not act would make a galaxy fly apart. In some galaxies, our own and Andromeda among them, a hundred billion stars move in nearly circular orbits in discs. In others, the less photogenic ellipticals, stars are swarming around in more random directions, each feeling the gravitational pull of all the others.

Galaxies are not as well understood as stars. Indeed as I shall explain, we do not even know what they are primarily made of. They interest cosmologists because they are probes, 'test particles', for structure and motions in the large scale Universe.

The nearest few thousand galaxies, those closer than about 300 million light years, have been mapped in both hemispheres. They are irregularly distributed, into clusters and superclusters. Are there, you may ask, clusters of clusters of clusters *ad infinitum*? There do not seem to be: our Universe is not, in modern jargon, a fractal. If it were, we would see conspicuous clumps in the sky however deep into space we probed. But the brightest *million* galaxies are actually fairly uniform over the sky; as we look at still fainter galaxies, probing still greater distances, clustering becomes less evident and the sky appears smoother.

There is in other words a well-defined sense in which the universe is roughly homogeneous. A terrestrial analogy may clarify this.

The ocean surface displays complex patterns: waves (sometimes small riding on large), foam etc. But once your gaze extends beyond the scale of the longest ocean swells, you see an overall uniformity, stretching to the horizon many miles away. A patch of ocean large enough to be 'typical' must obviously extend several times further than the scale of the longest waves. But this can still be small compared with the expanse of ocean we can see—our horizon extends far enough to encompass many patches statistically similar one to another, each large enough to constitute a 'fair sample'.

This broad-brush uniformity of seascapes is not, however, a general feature of landscapes: on land, progressively larger mountain peaks may stretch all the way to the horizon, and a single topographical feature may dominate the entire view.

Cosmology has only advanced because the volume out to the 'horizon' of our observation resembles a seascape rather than a mountain landscape. Even the largest superclusters are still small in comparison with the range of powerful telescopes. So we can define the average 'smoothed-out' properties of our observable Universe.

Expanding Universe

The overall motions in our Universe are simple too. Distant galaxies recede from us with a speed proportional to their distance, as though they all started off packed together 10–15 billion years ago.

Far away towards the horizon, we see domains whose light set out when the Universe was more compressed—more closely packed together. Astronomers can actually see the remote past—they have an an advantage over (for instance) palaeontologists, whose inferences depend on fossil traces.

The 'solid state' detectors used in modern telescopes (known as 'charged coupled devices', or CCDs) are fifty times more sensitive than photographic plates at detecting faint light. Telescope images reveal hundreds of thousands of faint smudges per square degree—each one of them a galaxy so far away that its light set out before our Solar System formed. These are typically only 2 or 3 arc seconds across. Do they look like nearby galaxies? Have they, for instance, acquired discs? Now that the Space Telescope works properly, it is yielding images sharp enough to answer this question.

Even more remote are the quasars, hyperactive centres of a special class of galaxies, so bright that they vastly outshine the 100 billion stars in their host galaxy. The 'distance record' is currently held by a quasar so redshifted that the wavelength of its light has been stretched, between emission and reception, by a factor 5.89. The Lyman-alpha 1216 A line, the strongest feature in the spectrum of hydrogen, reaches us in the red part of the spectrum, at around 7200 Å. This number, 5.89, is the factor by which the Universe has expanded since the light set out. These remote quasars tell us something important about cosmogony. When their light set out, when the universe was maybe only a tenth its present age, some galaxies (or at least their inner regions) had already formed, and runaway events in their centres had led to the extreme energetic activity that the quasar phenomenon manifests.

Quasars are probes of the era when galaxies were young, and perhaps just forming. But what about still earlier epochs?

Evidence for a 'big bang'

Did everything really start with a so-called 'big bang'? The idea goes back to Belgian Catholic priest Georges Lemaître in 1930. The phrase itself was introduced by Fred Hoyle, as a derisive description of a theory he never liked.

The name has stuck, and the clinching evidence for the theory came in 1965, when Penzias and Wilson found excess microwave noise, coming equally from all directions and with no obvious source, in their antenna at the Bell Telephone Laboratory. This has momentous implications—intergalactic space is not completely cold—it is about 3 degrees above absolute zero. That may not seem much, but it implies about a billion quanta of radiation—photons—for every atom in the Universe.

This 'cosmic background' causes some 1 per cent of the background 'fuzz' on a television set. It is an 'afterglow' of a pregalactic era when the entire Universe was hot, dense and opaque. After expanding for about half a million years the temperature fell below 3000 K; the primordial radiation then shifted into the infrared. The universe then literally entered a dark age, which persisted until the first stars in the first galaxies, and maybe also the first quasars, formed and lit space up again. The expansion has cooled and diluted the radiation, and stretched its wavelength. But this primordial heat is still around—it fills the Universe and has nowhere else to go!

We actually have firm grounds for believing that the temperature was once billions of degrees, not just thousands—hot enough for nuclear reactions. To explain why, I must recall the origin of carbon, oxygen and the other chemical elements.

In every galaxy, new stars are continually forming from gas clouds; other stars are ending their lives (sometimes violently as supernovae), and throwing material back into interstellar space. It is nuclear fusion that keeps stars shining. An essential byproduct of this is the synthesis of simple elements into other elements higher up the periodic table. Every galaxy is a kind of ecosystem whose gaseous content is being recycled through successive generations of stars. When our own Galaxy was young, there would have been no elements like carbon, oxygen and iron. Chemistry would then have been a dull subject. Before complex chemical compounds could form, and before a Solar System could emerge, ancient stars had to do the basic work of synthesizing, transmuting and recycling the chemical elements. This recycling led to the mix of chemical elements we see around us. We are the ashes of long-dead stars.

If the entire Universe had once been squeezed hotter than a star, you might wonder whether nuclear reactions could have happened then—some early proponents of the big bang theory suspected that the chemical elements were indeed forged in the early Universe. However, the expansion turns out to have been too fast to allow carbon, iron, etc. to be built up. But there would be enough time for about 25 per cent of the hydrogen to be turned into helium, and to make traces of deuterium and lithium.

What is remarkable is that the proportion of helium in old stars and nebulae, now pinned down with 1 per cent accuracy, turns out to be just about what is calculated. As a bonus, so are the proportions of lithium and deuterium. Moreover, these particular elements were a problem for the stellar nucleogenesis scenario that was so successful for carbon, oxygen, etc.: it was hard to explain why there was so much helium, even in the oldest objects; and deuterium is a fragile isotope that is destroyed rather than created in stars. These considerations therefore vindicate an extrapolation right back to when the Universe was hot enough for nuclear reactions to occur—that is when it was just a few seconds old.

How strong is the evidence for a big bang?

Over the last few years, the case for a 'big bang' has had several boosts—the COBE (cosmic background explorer) satellite showed that the background radiation had the expected 'black body' spectrum, to a precision of a part in 10 000; and cosmic helium and deuterium abundances have been measured more accurately. Moreover, there are several discoveries that might have been made, which would have invalidated the hypothesis, and which have not been made—the big bang has lived dangerously for 25 years, and survived.

The grounds for extrapolating back to the stage when the Universe had been expanding for a second (when the helium formed) deserve to be taken as seriously as, for instance, ideas about the early history of our Earth, which are based on inferences by geologists and palaeontologists which are equally indirect (and less quantitative)

There are some real true believers. The great Soviet cosmologist Zeldovich once claimed that the big bang was 'as certain as that the Earth goes round the Sun' (even though he must have known his compatriot Landau's dictum that cosmologists are 'often in error but never in doubt').

I would bet at least 90 per cent on the general concept, not quite 100 per cent. Consistency does not guarantee truth. Our satisfaction may be as illusory as that of a Ptolomaic astronomer who has just fitted a new epicycle.

The 'big bang' concept gives us the general framework in which we can (unless and until some glaring contradiction emerges) interpret the growing body of data, and address more detailed questions about how the Universe has evolved over its 10–15 billion year history.

You may be thinking: isn't it absurdly presumptuous to claim to know anything about the early stages of our entire observable Universe? Not necessarily. It is *complexity,* and not sheer size, that makes things hard to understand. In the primordial fireball everything must have been broken down into its simplest constituents. The early universe really could be less baffling, and more within our grasp, than the smallest living organism. It is biologists and the Darwinians who face the toughest challenge!

I shall come back later to our Universe's hot dense beginnings, and what happened in the first second, but let us now look forward rather than backward—as forecasters rather than fossil hunters.

Futurology

In about 5 billion years the Sun will die, and the Earth with it. At about the same time (give or take a billion years) the Andromeda Galaxy, already falling towards us, will crash into our own Milky Way and merge with it, forming a single amorphous elliptical galaxy.

Cosmic timespans extend at least as far into the future as into the past. Suppose America had existed for ever, and you were walking across it, starting on the East coast when the Earth formed, and ending up in California ten billion years later, when the Sun is about to die. To make this journey, you would have to take one step every two thousand years. All recorded history would be 3 or 4 steps. Moreover, these steps would come just before the half-way stage—somewhere in Kansas perhaps— not at the culmination of the journey!

In this perspective, we are still near the beginning of the evolutionary process. The progression towards diversity has much further to go. Even if life is now unique to the Earth, there is time for it to spread from here through the entire Galaxy, and even beyond.

But will the Universe go on expanding for ever, attaining some asymptotic heat death? Or will it, after an immense time, recollapse— the big crunch?

The ultra long range forecast depends on how much the cosmic expansion is decelerating. The deceleration comes about because every-thing in the universe exerts a gravitational pull on everything else. It is straightforward to calculate that the expansion will eventually go into reverse if the average cosmic density exceeds about 3 atoms per cubic metre. That does not sound much: but if the atoms in all the stars and gas in all the galaxies were spread uniformly through space, they would fall short of this 'critical' density by a factor of at least 50.

At first sight this seems to imply perpetual expansion, by a wide mar-gin. But it is not so straightforward. There seems to be at least 10 times as much material in 'dark' form as we see directly.

Dark matter

The discs of galaxies like our Milky Way or Andromeda contain neutral hydrogen gas, which does not itself weigh much, but serves as a tracer of the orbital motion. Radio astronomers can detect this gas, via its emission of the famous 21 cm spectral line. The gas extends far beyond the limit of the optically detectable disc. The orbital speed, inferred from the Doppler shift, is roughly the same all the way out. If the outermost clouds were feeling just the gravitational pull of what we can see, their speeds should fall off roughly as the square root of distance outside the optical limits of the galaxy; the outer gas would move slower, just as Neptune and Pluto orbit the Sun more slowly than the Earth does. So an extended invisible halo surrounds these galaxies—just as, if Pluto were moving as fast as the Earth, we would have to infer a heavy invisible shell outside the Earth's orbit but inside Pluto's.

Dark matter also pervades entire clusters of galaxies. The relative motions of the galaxies in a cluster can be measured, and they would disperse, rather than being bound within the cluster, unless their orbits were influenced by much more gravitating matter than we see. Other techniques (for instance X-ray detection of hot gas confined in the cluster) support this view. And there is now a new line of evidence—gravitational lensing. Richard Ellis and colleagues at the Institute of Astronomy have taken some remarkable sharp pictures of a cluster of galaxies, which reveal many faint streaks and arcs (Fig 1); these are remote galaxies, several times further away than the cluster itself, whose images are, as it were, viewed through a distorting lens. Just as a regular pattern on background wallpaper looks distorted when viewed through a curved sheet of glass, the gravity of the cluster of galaxies deflects the light rays passing through it.

Fig. 1 Gravitational lens in Galaxy Cluster Abell 2218. Hubble Space Telescope. Wide Field Planetary camera 2.

The visible galaxies in the cluster contain only a tenth as much material as is needed to produce these distorted images—evidence that clusters as well as in individual galaxies contain ten times as much mass as we see.

What could this dark matter be? Maybe it is faint stars whose centres are not squeezed hot enough to ignite their nuclear fuel. Or maybe black holes—remnants of big stars that were bright when the Galaxy was young but have now died.

Objects that do not radiate can still disclose themselves by their gravitational effect on light passing close to them. If a compact dark object moves across the line of sight to a background star, its gravity focuses the light ('gravitational lensing') so that the star appears brighter. The magnification rises to a peak (when the alignment is closest) and then declines again; so the background star brightens and fades in a predictable way. Unfortunately, this should not happen very often— alignment must be closer than 10^{-4} arc seconds to get the effect. Suppose we look at a background star, and ask what is the chance that there is a dark object along the line of sight to it, sufficiently well lined up to cause significant magnification. Even if there were enough compact objects to make up all the dark matter in our Galaxy, the chance is only about one in ten million.

To stand a chance of detecting an effect, either one must be prepared to wait for a very long time indeed or (more optimistically) one must observe not one but millions of background stars. Until recently, the data handling task would have been too daunting. But it is now feasible, and three independent groups have been monitoring, every clear night, several million stars in the small nearby galaxy (about 150 thousand light years away) known as the Large Magellanic Cloud.

We would expect these programmes to pick up thousands of variable stars of many types, and they have. Many astronomers make a living out of studying pulsating stars, flare stars, and binaries, but for this purpose the intrinsic 'variables' are a nuisance. The challenge is to pick out rare instances of variability that manifest the characteristic symmetrical rise and fall of a lensing event, and which are also achromatic, in the sense that the amplitude is the same in blue and red light. There are already two very impressive candidates for lensing events of the kind that would be caused by a 'brown dwarf' of around 0.1 solar masses. But the statistics are not yet good enough to settle what fraction of our Galaxy's dark matter could consist of these so-called 'brown dwarf' stars.

If, within the next year, several further putative lensing events are found that are equally convincing, it will imply that our Galaxy contains more gravitating stuff in brown dwarfs than in all the stars we see. Perhaps there will even be enough events to suggest that these faint stars constitute the dominant dark matter in our Galaxy. But if there are only a few further events (or if those recorded so far were lucky flukes) the quest for dark matter will have to continue by other means.

As a sociological digression, it is interesting that the scientists who embarked on these observing programmes had a background in particle physics—even though the techniques and instruments being used are of a primarily astronomical kind. Traditional astronomers were too easily discouraged by the daunting difficulties of monitoring millions of stars, and distinguishing lensing events from all the different kinds of intrinsically variable stars. But whatever the outcome, this technique has already been impressively vindicated.

Many physicists would be rather disappointed if all the dark matter were just low mass stars or black holes. They would be more excited if exotic particles were involved. For instance, neutrinos left over from the 'cosmic fireball' should be almost as abundant as photons: there would be just about a billion of them for every atom in the Universe. Even if they had a very tiny individual mass, their cumulative gravitational effects could be important. It is still controversial whether neutrinos have any mass at all—but if they do, it will have important consequences for dark matter. A recent claim from Los Alamos scientists would imply that neutrinos contributed more mass than the atoms we see, though still only a tenth of the critical density. But we should suspend judgement because the experimenters still disagree among themselves.

At least we know neutrinos exist. But theorists have a long shopping list of particles that might exist, and (if so) could have survived from the early phases of the big bang. These hypothetical particles, heavy but electrically neutral, would generally go straight through the Earth, just as neutrinos do. A tiny proportion, however, interact with an atom in the material they pass through, releasing a minuscule amount of energy. To detect these rare events—maybe one per day within every kilogram of material—you must go deep underground, to reduce other kinds of background, and many experimenters have taken up the challenge. (A mine at Bowlby in Yorkshire is being used by Peter Smith at the Rutherford Laboratory and his colleagues.) Even the optimists would not rate the chance of success for these underground experiments as being as much as evens. But the goal is still worth shooting for, because success would not only reveal a new class of elementary particle, but would also tell us what 90 per cent of the universe was made of—it would be at least as momentous a discovery as that of the microwave background in the 1960s.

The critical density?

The inferred dark matter in galaxies and clusters amounts to ten times what we see. But it still only adds up to about 20 per cent of the critical density.

There could, however, be some still more elusive material between clusters of galaxies. Its gravity would affect the motions of entire clusters of galaxies relative to each other. 'Streaming motions' of galaxies—deviations from the Hubble flow correlated on the scale of super-clusters—are now being studied. The gravitational pull implied by these motions seems so strong that it requires still more dark matter on the supercluster scale—perhaps even enough to supply the full critical density.

So much for the far future—what theologians call eschatology. If you are of an apocalyptic temperament and cannot wait a hundred billion years, then head for a black hole—you can there encounter a foretaste of the 'big crunch', created by a local gravitational collapse. Black holes form when heavy stars die, perhaps after some supernovae, and there may be many of them within our Milky Way. But, more spectacularly, monster black holes each weighing as much as a billion suns, the relic of the catastrophic event that formed a quasar, may lurk in the centre of some galaxies. You should aim, preferably, for one of the largest black holes. These are so capacious that, even after falling inside, you would have several hours for leisured observation before being torn apart in the middle. A more cautious course would be to remain in orbit just outside the hole. From that vantage point, if the hole were spinning fast, you would be safe, but would have a blueshifted and speeded-up preview of the future of the external Universe.

The dark matter could be anything from elementary particles up to black holes each weighing thousands of times as much as a star. The more of it there is, the more likely it is that exotic particles are impli-cated. That is because the proportions of helium and deuterium emerg-ing from the big bang depend on the density of ordinary atoms. The gratifying agreement I mentioned earlier would be lost if much more than a tenth of the critical density is composed of ordinary atoms. But this argument sets no limit on dark matter in neutrinos or other particles.

The challenge is to decide among many candidates for the dark matter whose gravity dominates our Universe. Our cosmic modesty may have to go a stage further. Copernicus dethroned the Earth from a central position. Hubble showed that the Sun was not in a special place. But now *particle chauvinism* may have to go. We ourselves, and all the stars and galaxies, would then be trace constituents of a Universe whose large scale structure is controlled by the gravity of dark matter of a quite different kind—we see, as it were, just the white foam on the wave-crests, not the massive waves themselves.

The evidence for dark matter comes mainly from applying Newton's inverse square law of gravity on scales where it cannot be checked independently. Some people have instead suggested that there is no dark matter but that Newton's laws may need modification. But it is hard to develop a theory that is even self-consistent, and jettisoning Newton, not just Einstein, is a high price to pay. Thus pre-Newtonian theories will remain my personal worst buy unless all possible dark matter candidates have been ruled out. That is far from the case now. Indeed the problem is to discriminate among an unduly long list of possible candidates.

There is certainly no strong evidence for there being enough dark matter to supply the full critical density. However, there is widespread theoretical prejudice (which I shall come back to later) that this should indeed be the case.

The 'age problem'

There is a different line of argument that may be a problem for this theoretically favoured option. That is the vexed question of how old the Universe is. We know how fast the galaxies are receding from us. If we also knew their distances, related to the so-called Hubble constant, we would know how much time had elapsed since the big bang if there were no deceleration—if the scale had increased linearly with time. The most recent Space Telescope data on cosmic distances are somewhat conflicting, but some astronomers claim that this time may be only 12 billion years.

But a Universe with the full critical density is being decelerated so much that it will eventually come to a halt. So its average expansion speed in the past was higher. The time since the big bang was only 2/3 of the Hubble time. That would mean only 8 billion years if this measurement were right (Fig. 2).

That seems in stark conflict with the estimated ages of the oldest clusters of stars. A star like the Sun can stay shining for 10 billion years. Heavier stars burn brighter and faster; stars smaller than the sun can live even longer. So if, for instance, a cluster contained stars of all masses up to that of the Sun, but none that were any heavier, we would infer a cluster age of 10 billion years.

In some clusters of stars, there do not seem to be any surviving

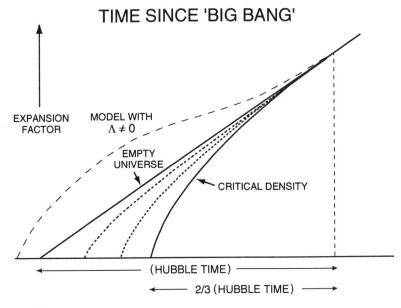

Fig. 2

members with more than about 0.8 of the Sun's mass. This implies that the cluster is 12 billion years old or even more. Since stars cannot be older than the Universe, a short Hubble time presents a big problem, especially for those who favour a 'critical density' for the Universe. Cosmic distance measurements involve a whole raft of interconnected uncertainties. An opinion poll among astronomers would give a wide range of answers for the 'Hubble constant'—the only point on which they would agree, I hope, is that it should not be settled by majority verdict, but will need more measurements and better understanding.

The 'error bars' on all measurements of the cosmic distance scale are still depressingly wide, even when the Hubble Space Telescope is used. If I were to place a bet now, it would be that the actual Hubble time will turn out to be long enough to make this discrepancy less stark. But it will be a challenge, and a more interesting one, if it does not.

Another possibility (indicated by the dashed line in Fig. 2) is that some other force (for example the cosmological repulsion force that can be added to Einstein's equations) causes a repulsion or acceleration.

It is embarrassing to admit that 90 per cent of the Universe is unaccounted for. But the existence of dark matter should not surprise us—there is no reason why everything in the Universe should shine. The problem is to distinguish among a long list of candidates. I am optimistic that, within five years, we will know what most of it is. There are three lines of attack.

1. Direct searches may detect some forms and eliminate others.
2. Particle physicists may some day be able to tell us the neutrino mass, and may develop firmer ideas on what other particles might survive from the ultra-early Universe.
3. Since large cosmic structures are gravitationally dominated by dark matter, the way they formed would plainly have depended on what that matter is and how it behaves. It may turn out, when structure formation is better understood, that the morphology of galaxies and clusters points strongly towards one particular option.

Emergence of cosmic structure

People often wonder how the Universe can have started off in thermal equilibrium—a hot dense fireball—and ended up manifestly far from equilibrium. Temperatures now range from blazing surfaces of stars (and their even hotter centres) to the night sky only 3 degrees above absolute zero. This seems contrary to thermodynamic intuitions that temperatures tend to equilibrate as things evolve, but it is actually a natural outcome of cosmic expansion, and the workings of gravity.

Gravity has the peculiar tendency to drive things further from equilibrium. When gravitating systems lose energy they get hotter. A star that loses energy and deflates ends up with a hotter centre than before. (To establish a new and more compact equilibrium where pressure can balance a (now stronger) gravitational force, the central temperature must *rise*.)

Gravity does something else too. It renders the expanding Universe unstable to the growth of structure, in the sense that even very slight initial irregularities would evolve into conspicuous density contrasts. Theorists are now carrying out increasingly elaborate computer simulations of how this happened. Slight fluctuations are 'fed in' at the start of the simulation: exactly how they are prescribed depends on the cosmological assumptions. The calculations can simulate a region containing a few thousand galaxies, large enough to be a fair sample of our Universe. As the expansion proceeds, regions slightly denser than average lag further and further behind. Eventually they stop expanding and condense into gaseous protogalaxies which fragment into stars. The same process on larger scales leads to clusters and superclusters

The aim is to make different assumptions about the initial fluctuations, the dark matter, etc., and see which leads to a pattern of structure closest to a typical sample of the real universe.

In the next few years, we can expect much better statistics on galaxies and clusters, not just in our own space–time neighbourhood, but (from observations of high-redshift objects) at earlier cosmic epochs. In parallel, computing techniques should become capable of incorporating greater physical realism.

Fluctuations post-COBE

Some recently discovered superclusters—'Great Walls' and other large-scale features in the galactic distribution—are so big that one may wonder whether these structures call into question the overall uniformity that is the basis for most of our thinking about cosmology. The simple 'homogeneous' cosmological models are still a good approximation. Even the largest superclusters span less than 2 per cent of the Hubble radius—a box that size is still large enough to be a 'fair sample' of the Universe.

A domain that is destined to become a supercluster must have an energy deficit. It must, for instance, be expanding a bit slower than average. The natural way to quantify this deficit is to compare it with the rest mass energy mc^2. This ratio is about one in 100 000. It is because this number is small that we can treat the Universe as approximately homogeneous, just as a sphere is nearly round if the height of the waves or ripples on its surface is only 1/100 000 of its radius.

If the Universe had started off completely uniform, it would still be so after 10 billion years. It would be cold and dull—no galaxies; therefore no stars, no chemical elements, no complexity, certainly no people. For cosmic structures to have 'emerged' via gravitational instability, then some kind of 'seed' fluctuations must have existed in the early Universe. The number, 10^5, which measures how big the fluctuations are, is one of the fundamental numbers characterizing our Universe.

The microwave background, a relic of the pregalactic era, should bear the imprint of these fluctuations. This radiation in effect comes from a very distant surface or horizon, at a redshift far beyond the quasars and a time long before the clusters had fully formed. Radiation from an incipient cluster on that surface would appear slightly cooler, because it loses extra energy climbing out of the gravitational pull of an overdense region. Conversely, radiation from the direction of an incipient void would be slightly hotter. The fractional differences in the temperature involve this same small ratio—they are only about one part in 100 000. This is a difficult target to shoot for.

Fluctuations with about the amplitude expected were first detected by NASA's COBE satellite. To measure such small effects was a technical triumph. But the fluctuations were not unexpected. It would have been far more baffling if they had not been there. The early Universe would then have been so smooth that it would not have been compatible with the conspicuous clustering we see in the present Universe, unless there were some process more efficient than gravity for moulding these structures and enhancing small initial contrast densities.

The early Universe was smooth in the same sense that the surface of the ocean is smooth—a well-defined mean curvature, with ripples on it. If you look down from the air on an ocean, you may first see just overall smoothness, But as your vision sharpens, you begin to discern some waves. A further modest improvement allows you to study wave statistics in detail. (Are the waves Gaussian? How does the amplitude depend on scale?) This is a metaphor for the exciting stage we are now entering in the study of the microwave background. COBE got the first positive results, but already these are being complemented and extended by ground-based and balloon experiments.

The fluctuations detected by COBE received disproportionate hype. Media attention is capricious, and when it reaches a certain threshold it feeds on itself—that certainly happened in this case. The fluctuations were important (even though not unexpected). But they were not even the most important result from COBE. That same satellite discovered that the background radiation had precisely a 'black body' spectrum—something nobody could have predicted confidently, and a measurement whose precision will not be surpassed for many years.

I am uneasy about how cosmology is sometimes popularized. If cosmologists claim too often to be 'stripping the last veil from the face of God', or making discoveries that 'overthrow all previous ideas', they will surely erode their credibility. It would be prudent, as well as seemly, to rein in the hyperbole a bit. (Otherwise journalists will have to become as sceptical in assessing scientific claims as they already are in assessing politicians.)

The very early Universe—the first millisecond

When the Universe was a second old (when the helium was made) the matter was no denser than air; conventional laboratory physics is applicable, and is vindicated by the agreement with observations. But the further we extrapolate back, the less confidence we have that known physics is either adequate or applicable. For the first millisecond, everything would have been squeezed denser that an atomic nucleus. For the first 10^{-14} seconds the energy of every particle would surpass what even CERN's new accelerator will reach.

Be warned that I am now entering speculative territory, where even Zeldovich would harbour some doubts.

First, what about the initial expansion rate? This has to be very precisely tuned. The two eschatologies—perpetual expansion or recollapse to a 'crunch'—seem very different. But our Universe is still expanding after 10 billion years. Had it recollapsed sooner, there would not have been time for stars to evolve—indeed, if it had collapsed after less than a million years it would have remained opaque, precluding any thermodynamic disequilibrium. On the other hand, the expansion cannot be too much faster than the critical rate. Otherwise gravity would have been overwhelmed by kinetic energy and the clouds that developed into galaxies would have been unable to condense out.

Inflation

The initial expansion seems to have been set up in a rather special way. In Newtonian terms the initial potential and kinetic energies were very closely matched. How did this come about? And why does the Universe have the large-scale uniformity which is a prerequisite for progress in cosmology?

The answer may lie in something remarkable that happened during the first 10^{-36} seconds, when our entire observable Universe was compressed in scale by 27 powers of ten (and hotter by a similar factor). Ever since that time, the cosmic expansion has been decelerating, because of the gravitational pull that each part of the Universe exerts on everything else. But theoretical physicists have come up with serious (though still, of course, tentative) reasons why, at the colossal densities before that time, a new kind of 'cosmical repulsion' might come into play and overwhelm 'ordinary' gravity. The expansion of the ultra-early Universe would then have been exponentially accelerated, so that an embryo Universe could have inflated, homogenized, and established the 'fine tuned' balance between gravitational and kinetic energy when it was only 10^{-36} seconds old.

This generic idea that the Universe went through a so-called inflationary phase is compellingly attractive. The fluctuations from which clusters and superclusters form, and the even vaster ones whose imprint on the background radiation spreads right across the sky, may be the outcome of microscopic quantum fluctuations when everything we can now see was squeezed smaller than a golfball. We do not of course know the physics that prevailed at this ultra-early time. But there is a real prospect of discovering something about it. Specific models of how the inflation is driven make distinctive predictions about things we can observe—large scale clustering, and small non-uniformities in the background radiation over the sky. We shall soon be confronting the inflationary era of cosmic expansion with real empirical tests, just as we can already, by measuring the abundances of helium and deuterium, learn about physical conditions during the first few seconds.

The inflationary idea also, incidentally, strongly suggests that the mean cosmic density is very close to the 'critical' value that demarcates the boundary between perpetual expansion and eventual recollapse— that is the basis of the prejudice I mentioned early in favour of the critical density.

The Universe then has, in a sense, zero net energy. Every atom has an energy because of its mass—Einstein's mc^2. But it also has a negative potential energy due to the gravitational field of everything else, and this could exactly balance its rest mass. Thus it may cost nothing, as it were, to expand the mass and energy in our Universe.

Physicists sometimes loosely express such ideas by saying that the Universe can essentially arise 'from nothing'. But they should watch their language, especially when talking to philosophers. The physicist's vacuum has all particles and forces latent in it—it is a far richer construct than the philosopher's 'nothing'.

Any theory of the 'beginning' of our Universe would, of course, be hard to check, and may never be taken too seriously unless it has a compelling inevitability about it—a resounding ring of truth that compels assent. And it in any case would not tell us *why* there were a universe. To quote Stephen Hawking: 'What is it that breathes fire into the equations? Why does the Universe go to all the bother of existing?'

Exotic relics

The structures in the Universe, and the atoms themselves, are fossils of an ultra-early era. So are the atoms, or their constituent quarks. Some other remarkable fossils have been conjectured by theorists, and are being seriously looked for—magnetic monopoles; or small black holes the size of an atom, but weighing as much as a mountain. Even more astonishing are cosmic strings—vast elastic loops, thinner than an elementary particle, but long enough to stretch across the universe, flailing around at nearly the speed of light, and heavy enough for their gravity to affect entire galaxies.

If any of these were discovered, they would be crucial links between the cosmos and the microworld. In the ultra-early Universe, the mysteries of the cosmos and the microworld overlap. Processes that occurred as early as 10^{-36} seconds may have imprinted the excess of matter over antimatter, the ripples in the fabric of space–time, and perhaps the physical laws themselves. Our entire Universe may even be part of a still grander ensemble (Fig. 3).

Concluding assessment

Let me conclude by trying to assess the state of play—where we can lay confident bets and where we should not (or not yet).

It is helpful to divide cosmic history into three parts (Fig. 4).

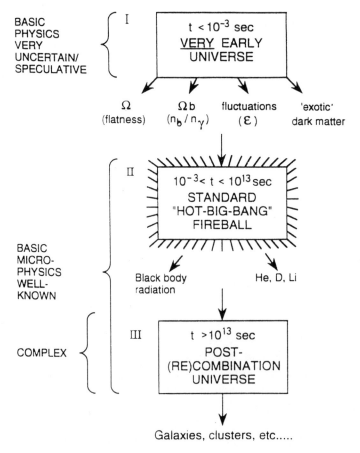

Fig. 3

Part 1 is the first millisecond, a brief but eventful era spanning forty decades of logarithmic time, starting at the Planck time (10^{-43} seconds). This is the intellectual habitat of the high energy theorist and the 'inflationary' or quantum cosmologist.

The second stage runs from a millisecond to about a million years. It is an era where cautious empiricists like myself feel more at home. The densities are far below nuclear density, but everything is still expanding in an almost homogeneous fashion. The relevant physics is firmly based on laboratory tests, and theory is corroborated by good quantitative evidence—the cosmic helium abundance, the background radiation, etc. Part 2 of cosmic history, though it lies in the remote past, is the easiest to understand.

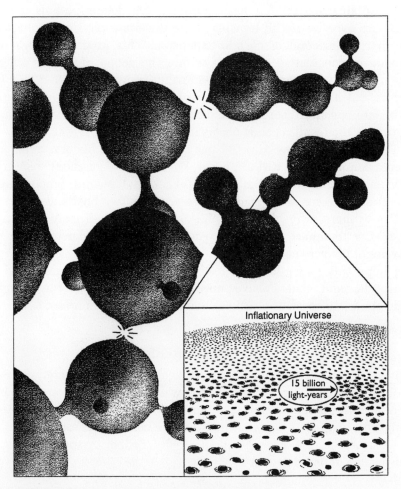

Fig. 4 Ensemble of self-reproducing universes.

The tractability lasts only so long as the Universe remains amorphous and structureless. When the first gravitationally bound structures condense out—when the first stars, galaxies and quasars have formed and lit up—the era studied by traditional astronomers begins. We then witness complex manifestations of well-known basic laws. Gravity, gas dynamics, and feedback effects from early stars combine to initiate the complexities we see around us and are part of. Part 3 of cosmic history is difficult for the same reason that all environmental sciences—from meterology to ecology—are difficult.

Cosmology has progressed because the laws of physics we study in the laboratory apply in the remotest quasar, and back to the first few seconds of the 'big bang'. When that is not so—when there is not a firm link with laboratory science—cosmologists are on shakier ground. But we are now realizing that the few basic numbers that determine how the Universe has evolved are all legacies of the uncertain physics of phase 1— the first millisecond of cosmic expansion. This progress brings some previously speculative questions within the scope of serious science. The ultra-early Universe is still an arena for controversy rather than consensus, but it is now accessible to real scientific investigation—we can formulate specific theories, calculate their cosmological consequences, and check the results against observation.

Telescopes can now view 90 per cent of cosmic history; other techniques can probe still earlier phases. We are confident about cosmic history—at least in very broad outline—back to 1 second, when the first chemical elements were made. There are now two challenges. First, to delineate more fully how a ten-billion-degree fireball evolved into the cosmos astronomers now study. And, second, to understand the new physics of the very earliest stages.

I hope this paper has given some flavour of the progress and the prospects, and that I have not transgressed Niels Bohr's famous injunction—to speak as clearly as one thinks, but no more so.

MARTIN REES

Martin Rees is a Fellow of King's College and a Royal Society Research Professor at Cambridge University. He was formerly Director of the Institute of Astronomy at Cambridge, and before that a Professor at Sussex University. His main research interests are cosmology, astrophysics and space research. He was President of the British Association for the Advancement of Science in 1994–95.

Nuclear power plant safety— what's the problem?

JOHN G. COLLIER

Introduction

Nuclear power is a clean, economic but controversial method of generating electricity. There is public apprehension because of the early relationship with nuclear weapons and, therefore, the association with radiation and radioactivity. Even so, over 430 nuclear plants worldwide currently supply 17 per cent of world electricity demand (*see* Figure 1). The figure is over 30 per cent in Europe and around 25 per cent in England and Wales.

Nuclear power could make an even greater contribution in the future. Global population will continue its increase and with it energy demand. Fossil fuels are finite. Their increased use will cause environmental

Nuclear share (%)

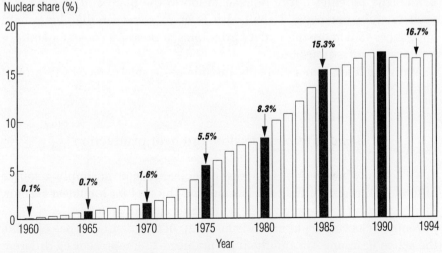

Fig. 1 Graph of global nuclear electricity generation since 1960.

problems. This will challenge the sustainable development of the planet.

However desirable that greater contribution from nuclear power, it will not materialize without wider public acceptance. Four concerns are highlighted time and time again. They are

> leaks of radioactive materials from nuclear plants,
>
> the decommissioning of old power stations, coupled with the disposal of radioactive waste,
>
> and finally the safety of nuclear power stations.

The last of these—the safety of nuclear power stations—is the subject of my discourse, and I aim to dispel some of the mystery surrounding the topic.

The safety of nuclear plant is based on well understood physical phenomena as well as sound engineering practice.

The basic principles of reactor safety can be referred to as the three 'Cs':

> CONTROL the reaction,
> COOL the fuel,
> CONTAIN the radioactivity.

The other essential ingredient is a dedicated, well trained and educated operational staff. This all adds up to DEFENCE IN DEPTH.

To illustrate these matters, in what follows, I shall draw on the capability of a simulator which represents or models a nuclear power plant. This provides a true representation of the plant and as such it is used to train operators. I shall also refer to some familiar household appliances to demonstrate safety principles: a pressure cooker, a kettle, a hair dryer, and a mousetrap.

Many such appliances already include basic safety devices to protect us all.

Basis of nuclear fission and heat production

For electricity generation purposes, a nuclear reactor is simply a source of heat. The origin of the heat is uranium, the fuel for a nuclear reactor. Figure 2 shows the form of oxide pellet used in many reactors.

Naturally occurring uranium is made up of two distinct types of atom, the active uranium-235 and the inert uranium-238, representing different atomic weights. The active uranium-235 type is present only in small

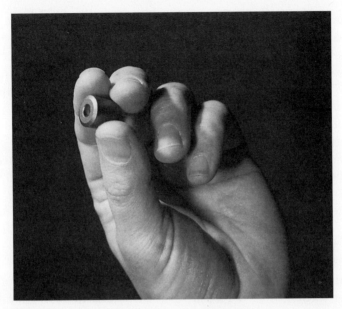

Fig. 2 Photograph of a uranium dioxide pellet.

amounts—seven parts in one thousand, or 0.7 per cent. Uranium-235 is unique in that, if a slow neutron collides with the nucleus, it will split into two new smaller nuclei, releasing the binding energy (*see* Figure 3.) The process is known as fission and the new atoms as 'fission products'.

As an energy source, each kilogram of uranium-235 fissioned releases the equivalent of 3 million kilograms of coal burnt. The fission process

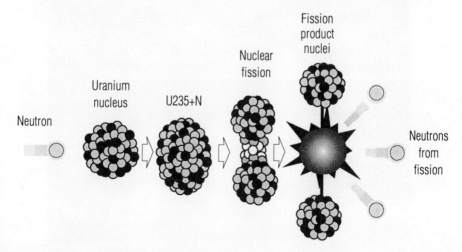

Fig. 3 Diagram of uranium fission.

also releases several more neutrons. These, in turn, can cause further uranium-235 atoms to split.

This behaviour can be illustrated with a simple model of a uranium-235 atom—a modified mousetrap (*see* Figure 4). The neutrons are represented by little plastic balls. If a 'neutron' lands on the mousetrap it will release energy and two more 'neutrons'.

If, through this multiplication of neutrons and successive fissions, a self-sustaining 'chain' reaction can be induced, it can be used to produce heat to generate electricity. Alternatively, if the same energy release can be contrived to be very rapid, it can be used to produce an atomic weapon.

However, because neutrons are lost from the surface or captured by the inert uranium-238, it is impossible to produce a self-sustaining sequence with natural uranium by itself. So we need to improve the efficiency of the uranium-235–neutron interaction.

Neutrons produced by fission have a very high velocity—20 000 kilometres per second. If their velocity is reduced considerably—by a factor

Fig. 4 Photograph of modified mousetrap with ball held above.

of 10 000 to two kilometres per second—they are then one thousand times more likely to induce fission. This slowing down can be achieved by multiple collisions—say around 100—with light atoms like carbon or hydrogen (a process known as 'moderation').

If carbon, in the form of graphite, is used as the 'moderator' then the improved efficiency of the process allows a chain reaction to be induced using naturally occurring uranium. This is the basis of the Magnox reactor. If hydrogen is used as the moderator—in the form of water—then a chain reaction still cannot be induced with natural uranium. This is because the water also captures some of the neutrons. To induce a chain reaction with water moderation, as in light water reactors, the proportion of the active uranium-235 in the fuel needs to be increased to around 4 per cent, by a process known as enrichment.

The self-sustaining chain reaction can be demonstrated, this time by a collection of mousetraps. Triggering off one mousetrap sets off the chain reaction (*see* Figure 5). Typically, in a graphite moderated reactor, the time period between fissions is one thousandth of a second. Alternatively, to

Fig. 5 Photograph of mousetraps—with all the balls in the air.

make a weapon it is necessary to separate the active uranium-235 from the inert uranium-238. This active uranium-235 is then used unmoderated. So the time interval between fissions is then less than one millionth of a second. In a nuclear reactor, the presence of both the inert uranium-238 as an absorber of neutrons and the moderator, which induces neutron lifetimes 1000 times longer between fissions, ensures that a nuclear reactor cannot possibly explode like a nuclear weapon.

Let us turn now to the means by which controlled nuclear chain reactions are used safely in electricity generating plant.

How reactors work

Looking at the chain reaction in a little more detail, the various fission processes can be divided into three categories:

(1) Those immediate processes related to the fission reaction, the so-called prompt processes resulting in the production of
fission products,
heat arising from their kinetic energy,
gamma radiation, and
two new neutrons emitted.

(2) Those processes which effectively lose neutrons from the system:
the capturing of neutrons in inert uranium-238 to form plutonium-239, which itself fissions if a neutron collides with it;
the parasitic absorption of a neutron in an atom of structural material;
leakage from the reactor assembly.

(3) Delayed processes arising from the decay of fission products:
first, beta and gamma radiation from fission products together with the associated heat release;
'delayed' neutrons emitted from fission products.

So what is the overall balance of these processes in an operating reactor?

Figure 6 illustrates the fate of neutrons in an advanced gas cooled reactor (AGR) in steady operation. Here, 100 fissions produce, on average, 259 neutrons. Fifty-nine of these neutrons are captured by the reactor structure or by fission products or escape from the reactor assembly. A hundred neutrons are absorbed by the uranium fuel, mainly uranium-

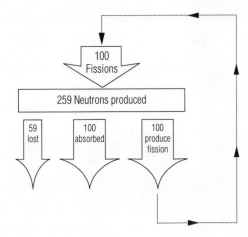

Fig. 6 Diagram of neutron fate.

238, producing the new material, plutonium-239, itself capable of fission. The remaining hundred neutrons cause the further fissions necessary to sustain the reaction.

We thus have a delicate balance from which a slight deviation would cause the chain reaction either to die away or to accelerate. Fortunately, nature has provided inherent features of nuclear reactions which ensure that changes to this balance are not so rapid that they cannot be properly controlled.

To understand this we need to distinguish between the 'prompt' processes (fission and instantaneous release of heat and production of fission neutrons) and the 'delayed' processes (production of energy and neutrons from decay of the fission products). These delayed effects are very important. The delayed neutrons (7 parts in a thousand of the total) play a vital part in the reactor control and safety systems. Also, heat continues to be generated after the fission reaction is stopped. It must be taken care of by the reactor cooling systems.

The rate of decay heat production falls quite rapidly after shutdown of the fission reaction but is nevertheless very significant. For example, an AGR generating 1500 MW of thermal energy will be producing 24 MW one hour after shutdown (*see* Figure 7). It will still be producing 5 MW ten days after shutdown. Removal of this decay heat, or 'after'-heat, is a very important consideration in reactor safety.

Basic components of a nuclear reactor

Now let us turn to the components of a nuclear power station. In doing so we will begin to think about the three Cs: control, cool and contain. I

(Power in MW corresponds to a 1500 MW Th AGR)

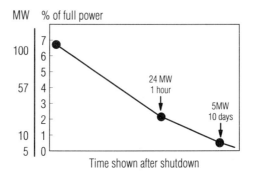

Fig. 7 Diagram of the production of decay heat following shutdown.

have chosen the advanced gas cooled reactor (AGR) for most of my illustrations (*see* Figure 8), though I will refer to water cooled reactors a little later on.

First let us look at the COOLING. In the case of an AGR, carbon dioxide at high pressure is driven over the fuel by large fans (circulators). The fuel consists of pellets of slightly enriched uranium in oxide form sealed in a can made of stainless steel (Fig. 9). The fuel assemblies are

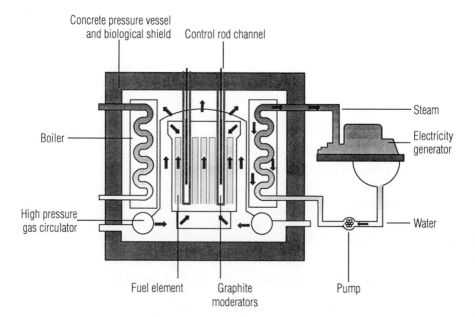

Fig. 8 Schematic diagram of an AGR.

Fig. 9 Photograph of a fuel pin.

stacked in vertical holes (channels) in the massive structure of the graphite moderator.

Figure 8 shows the whole is contained in a pre-stressed concrete pressure vessel, retaining the high pressure carbon dioxide gas. The carbon dioxide extracts heat from the fuel elements. This heat is then used in a boiler or steam generator to convert water to steam. The steam is then passed through the turbine, which drives the electrical generator. The very low pressure exhaust steam from the turbine is passed to a condenser, where it is converted back into water and fed back to the steam generator.

Thus the 'COOL the fuel' principle is met by maintaining the gas and water circulation systems in operation. It is important to match the cooling to the heat generated by the fuel so that safe temperature limits are not exceeded.

Next, the means of 'CONTROLLING the reaction'. For this, we need to manage the neutron population. This is done by introducing special materials such as boron which readily absorb neutrons.

Control rods, which incorporate absorber material, are driven in or out of the reactor according to whether the neutron population and therefore the rate of fission is to be maintained, increased or diminished. The number of fissions determines the rate of heat production. So this offers the means of changing the power output of the reactor. The rods have two different functions. They control the neutron population and hence the reactor power (as part of the CONTROL system). But, if necessary, they can drop into the core very rapidly to ensure that the chain reaction ceases altogether (as part of the reactor protection system).

I now return to those specific features of the fission process which allow us to control the reaction. As we have seen, most of the neutrons present in a reactor are the so-called prompt neutrons from the fission process. In graphite moderated reactors they have a lifetime of typically one thousandth of a second. If the neutron population consisted only of these prompt neutrons, the reactor would be very difficult to control. Rapid changes in neutron population would occur.

Fortunately, the 7 out of a thousand of the neutrons, those which arise from the decay of fission products (rather than directly from the fission process itself), are produced after delays ranging from a fraction of a second up to 80 seconds. So in a reactor at steady state, the prompt neutrons are 'topped up' by these delayed neutrons. Essentially the control and safety shutdown systems operate on these delayed neutrons. The delayed neutrons increase the effective neutron lifetime in the reactor by a factor of 100, to one-tenth of a second.

The effect of this is a very dramatic damping down of the rate of change of neutron population. Figure 10 shows this effect by comparison of the variation of neutron population in an AGR plant calculated for two cases: (a) all neutrons are 'prompt', and (b) the normal population of delayed neutrons is included.

The effect of the delayed neutrons is such that control rod movement responses on a time scale of 10–20 seconds readily control the process. Delayed neutrons thus play a vital role in the control of nuclear reactors.

Another important process relates to the stability of the reaction as temperature conditions change. An increase in fuel temperature causes increased capture of neutrons by uranium-238 and a consequent decrease in the proportion of neutrons available to provide fission. So the neutron population decreases and the reactor power falls. This provides a self-stabilizing, or self-regulation, effect because an untoward increase in fuel temperature is partially self-compensated by a reduction in reactor power. This greatly eases the task of the automatic control and shutdown systems.

So the 'CONTROL the reaction' safety principle involves ensuring effective management of the chain reaction using neutron absorbing

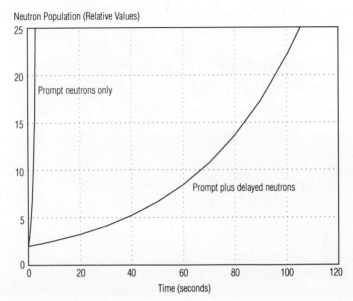

Fig. 10 Graph showing the effect of delayed neutrons in limiting the rate of power increase.

material. This is made possible by the basic phenomena of the delayed neutrons and the self-stabilizing effects of fuel temperature changes.

Finally we come to the last principle 'CONTAIN the radioactivity'.

The nuclear fission process results in intense radiation. The fission products also continue to emit radiation after the fission reaction is closed down. So it is very important to provide proper shielding around the reactor. As Figure 8 shows, this shielding takes the form of a thick concrete 'biological shield'. In AGR plant, the pre-stressed concrete pressure vessel doubles as the biological shield. Where necessary, as it is for water reactor plant, further protection is provided by housing the whole system inside a leak-tight containment building. Figure 11 shows the containment for Sizewell B under construction and illustrates the double containment structure. These are the components in place specifically for containment of the radioactivity. However, protection of people requires multiple barriers. These are illustrated in Figure 12.

Most fission products are retained within the fuel—the first barrier. The fuel is sealed in metal tubes, stainless steel for AGRs, zirconium alloy for water reactors. This is strong enough in all normal circumstances to contain all the fission products which escape from the fuel itself and is a second barrier. The reactor pressure vessel which contains the core and the high pressure coolant forms a third barrier. And for water reactors there is the further barrier of the containment building.

Fig. 11 Photograph of Sizewell B containment under construction.

The whole purpose of the safety systems is to ensure that these barriers are not challenged and all remain intact. The safety limits are defined with this specific objective in mind. So 'CONTAIN the radioactivity' is met by a multi-barrier approach. And safety systems are in place with the objective of preventing these barriers being breached.

The simulator (Demonstration 1)

Figure 13 shows a reactor simulator which can be used to demonstrate some of the principles I have outlined.

In the centre of the display is a mimic of an AGR. Panels show a number of important operational parameters:

> the temperatures of the carbon dioxide coolant where it enters and exits from the reactor;
>
> the position of the control;
>
> reactor gas pressure;
>
> reactor power;
>
> indication of the rate of change of the neutron population;

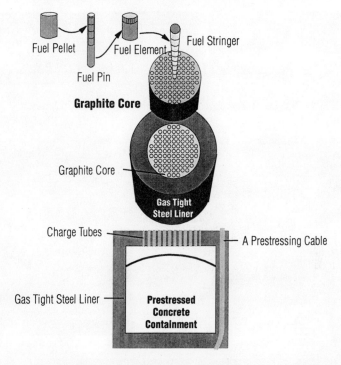

Fig. 12 Schematic diagram of the multi-barrier containment system for an ACR.

maximum fuel can temperature (calculated);

indicators of the various signals from sensors ensuring automatic reactor shutdown known as a reactor 'trip'.

A VDU display shows the time variation of many of these parameters. Note that the representation is entirely realistic. The display **is** simplified but it does give essential control room data as would be displayed to the AGR station operators.

Since I now intend to discuss faults and breakdowns, I need to make an important general comment. The idea that we give consideration to faults at all may cause apprehension. It should not. The whole approach is to assume that faults will occur and then build in defences to cope with such faults. So the practice, which I am going to illustrate, of examining the likelihood and consequences of faults, sometimes very unlikely severe events, must be seen as a source of confidence—not of expectation that harmful accidents may occur.

Figure 14 is an illustration to demonstrate the self-regulation effect of fuel temperature increases which I briefly touched on earlier. It shows the time variation of (a) reactor gas flow, (b) reactor gas outlet tempera-

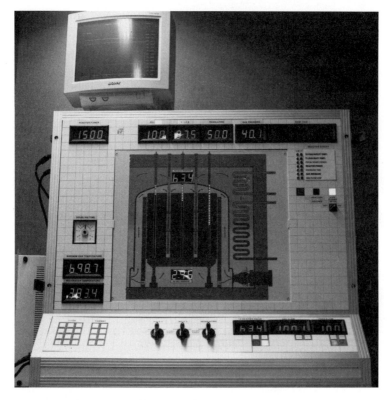

Fig. 13 Photograph of the simulator.

ture (T_2), (c) reactor gas inlet temperature (T_1), and (d) reactor power, taken from the simulator VDU display.

A very severe fault was introduced to increase fuel temperatures. This was done by stopping half the circulators, greatly reducing the flow of gas over the fuel. The reactor control system was also inhibited so that the inherent stabilizing effect could be observed more easily.

In steady operation the reactor parameters displayed were steady. When the fault was initiated at time t_0, the gas flow decreased sharply. This increased reactor temperatures generally, including reactor gas temperatures T_1 and T_2 and fuel temperature. In consequence, the neutron multiplication rate was reduced and so the reactor power fell by around 20 per cent at $(t_0 + 20)$ seconds. Note there was no correcting movement of the control rods, because the control system was inhibited, so the power reduction shown was inherent, a result only of the increase in fuel temperature.

To recapitulate, the reactor gas flow was halved, gas temperatures (and fuel temperatures) increased, and the reactor power fell.

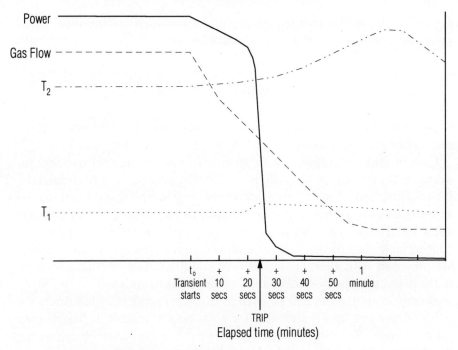

Fig. 14 Graph of self-regulation.

The effect of increasing fuel temperatures that might have been expected (a power increase) was countered by the self-regulating effect which I previously described.

You may wonder at my emphasis on what seems to be a minor trend on a graph, but it is far from minor in its importance. Without this self-stabilizing trend, reactor control would be extremely difficult, if not impossible. In fact, the Chernobyl accident came about in considerable measure because of the unique and highly unsatisfactory design features of the Russian plant and of the operating mode on the day. These combined to override and overwhelm this self-regulating effect of fuel temperature. As a result an uncontrollable power increase occurred which destroyed the plant.

Let us consider the simulation further. Self-regulation notwithstanding, plant temperatures (T_1 and T_2) continued to rise, since only half the gas circulators were operating. There are, of course, temperatures beyond which safety might be prejudiced. The reactor protection system therefore detected the temperature increases and automatically terminated the reaction, triggering an audible alarm and tripping the reactor by causing the control rods to drop into the core reactor when the gas outlet temperature reached its pre-defined safe limit (TRIP at ($t_0 + 24$) seconds).

Figure 14 shows the post-trip power and temperature reductions. I will return later to features of the shutdown systems.

Hair dryers, kettles and reactor cooling

I turn now to some very well-known simple concepts and their application in nuclear plant safety.

Certain faults can give rise to an interruption in normal cooling: for example those related to coolant circulation pumps or gas circulators. When such an interruption occurs, the reactor is shut down by its automatic safety systems. But, as we discovered earlier, after-heat generation continues after shutdown of the fission reaction owing to the continuing decay of the fission products formed. So all reactor systems are provided with alternative means of cooling to remove this after-heat in the event that the normal cooling system fails to operate.

Domestic appliances can be used to illustrate some simple principles. A hair dryer is a useful analogue for a gas cooled reactor. It has its own safety system. To cool the heating element the fan must be operating. Block the air intake, by cupping one's hand over the end, for example, or disable the fan, and a temperature sensor inside detects overheating and switches off the power supply. So too in AGRs. Loss of circulation is detected by sensors which shut down the reactor.

An important feature of a gas cooled reactor is that, even if all circulators in a pressurized reactor were to fail, overheating is prevented by the well known phenomenon of natural convection. The point can be illustrated simply using the apparatus in Figure 15. This is a hair dryer modified so that the power to the heating element and to the fan can be controlled separately and with a sensor measuring the temperature close to the heater windings.

With the dryer in normal operation, the heater temperature was about 550°C. To simulate the tripping of the reactor followed by the decay heating, the fan was stopped and the power was reduced to 10 per cent. Figure 16 shows the result. The temperature first rose a little, then fell as natural convection through the heater automatically took charge, the windmill in Figure 15 illustrating the induced air current. To emphasize the importance of the induced air current, a chimney extension was used to enhance the cooling, as was shown by increased windmill speed. The temperature fell further, eventually to less than 300°C, and the windmill speeded up.

This illustration is not just a reminder of the significance of chimneys. Rather it is to emphasize that we use the well-known effect of natural

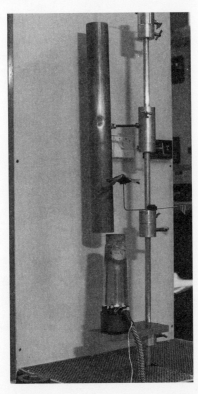

Fig. 15 Photograph of a hair dryer with windmill and chimney.

convection, a gravity driven phenomenon, to provide inherent safety. In the very unlikely event that all circulators fail on an AGR, the tripping of the reactor would be followed automatically by establishment of such a natural convection regime. This would transfer decay or 'after'-heat very effectively from the fuel to the boilers.

Water reactors

So far I have talked about gas-cooled reactors. Recently we started up our first water reactor at Sizewell (Figure 17). In such reactors, the coolant and the moderator are both ordinary water. Though very different in design from gas-cooled reactors, some features are still common. For example, when the fission reaction has been stopped by shutting the reactor down, decay heat continues to be generated. As with the AGR, special arrangements are needed if there are problems with normal cooling systems.

A break in the main reactor pipework, while very unlikely, would

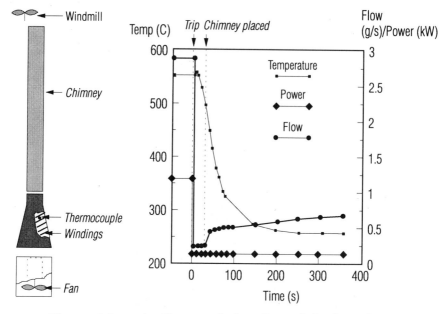

Fig. 16 Schematic diagram of the effect of the hair dryer experiment.

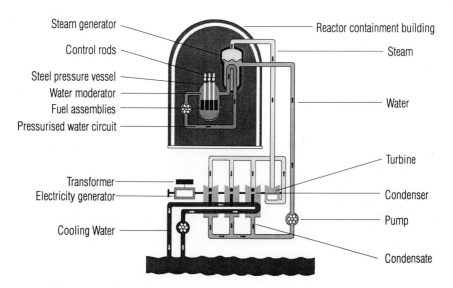

Fig. 17 Schematic diagram of a pressurized water reactor nuclear power plant.

cause the loss of supply of the water cooling the fuel. Although this would terminate the fission reaction, because the water is also the moderator, this is one of the more extreme plant states that has to be catered for. An analogue for this is a kettle.

Tea-making encompasses several different plant states: filling, heating up, boiling, and pouring.

If we leave the kettle element switched on, the last of these is equivalent to a loss of reactor coolant with continued decay heat generation.

The operational parameters in the case of the reactor simulator are listed above. The equivalents for the kettle are

the amount of water present in the kettle—the inventory,

the power input to the element,

the temperature of the water,

the surface temperature of the element, and

the temperature of the element wiring.

We are specially interested in the last two. These relate to the safety limits not to be exceeded if the heater is not to be damaged.

If a kettle is emptied with the heater still on, the heater surface temperature will increase sharply. If the kettle is returned to the horizontal, the heater will be quenched by the water remaining. The effect is marked by the audible hiss we have all experienced. This emphasizes that, while the power is on, the heater can be protected by keeping it covered with water and getting rid of the heat by boiling. What has this to do with water reactors?

A kettle has many of the characteristics of a water reactor system. The element in a kettle is of similar dimensions to that of a PWR fuel pin (see Figure 18). Indeed the power generated per unit length of an electric kettle element (approximately 2 kW m) is similar to the decay heat in an equivalent length of PWR fuel pin immediately following shutdown. As with the kettle, the reactor has both inherent and engineered safety features which terminate the reaction if there is loss of coolant. What is then needed to remove the decay heat is to keep the fuel covered by water. This is done by injecting an emergency water supply. For the kettle, this means simply topping up from the tap. For a PWR it means a special supply of water injected into the primary circuit of the reactor to make up any water lost (Figure 19).

Let me emphasize the importance of this further by reminding you that, at Three Mile Island in 1979, for two hours, this requirement for emergency cooling was not met. So the fuel overheated and suffered gross damage. Large quantities of fission products escaped—but only

Fig. 18 Photograph of a kettle element and fuel pin.

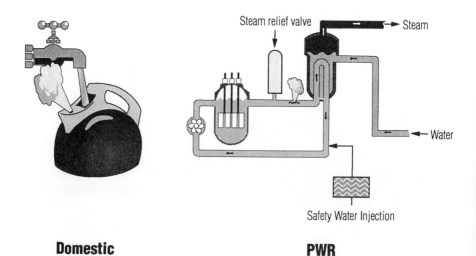

Domestic   **PWR**

Fig. 19 Diagram of the two kettles.

into the reactor vessel and containment building. The containment did
its job well.

Let me recapitulate. We have demonstrated how the basic reactor
physics of delayed neutrons and self regulation provide inherent safety.
We have seen how the first of the three Cs, Control, is exercised by the
simple process of adjusting the position of neutron absorbing material.
We have demonstrated how the decay heat can be removed in a gas-
cooled reactor by a simple Cooling process, natural convection, and in a
water reactor by keeping the core covered with water. In passing we
have also seen how gravity is used to advantage in the shutdown and
natural convection systems.

But these simple processes and fundamental features need to be
brought into play when required. This is done by the action of sensors
located on key plant items to indicate when an unsatisfactory condition
is being approached. They then initiate the action of the safety systems.
This total Reactor Protection System has to be highly reliable.

A pressure cooker serves as an analogue to show some of the means
by which reliability is achieved (see Figs 20 and 21). This was adapted
by providing an electric heating element of the same type used in an
electric kettle ((a) in Fig. 20) with an electricity supply controlled
through a transformer. Boiling water in the cooker produces steam at (b),
which is fed via a jet to a small turbine generator to make electricity.

In performing experiments, so as to avoid blowing up the cooker
or melting the heater, two safety limits are imposed, the cooker failure
pressure and the heater melting temperature.

The cooker is well equipped. It has a pressure relief valve (c) and the

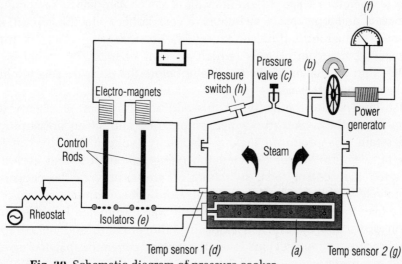

Fig. 20 Schematic diagram of pressure cooker.

Fig. 21 Photograph of a pressure cooker.

supply plug is the sort that will eject if the element overheats. These provide protection. However, if the heat were nuclear in origin there would be a need to cater for the unexpected and for high reliability. The means by which this can be done may be shown by some simple experiments.

A temperature sensor (d) on the wall of the cooker detects temperature increases and operates a switch in a low voltage electric circuit if a defined temperature limit is exceeded. Breaking this electric circuit actuates some 'control rods' which in turn operate the switches (e) which shut off the supply to the heater before the pressure gets too high or the element overheats.

To demonstrate the action of this additional system, electricity was supplied to the cooker, the water temperature and steam pressure rose and steam was produced making electricity (voltage displayed at (f)). The temperature sensor detected the rising temperature and at 102 °C operated the control rods, switching off the supply to the cooker as planned. An audible warning and indicator light confirmed the 'trip'.

To take the experiment further, a fault was introduced into this automatic protection system. On restarting, the pressure and temperature increased above the previous level, the turbine generator speeded up and a higher current was produced.

At a temperature of 103 °C a duplicate temperature sensor (g) providing a back-up tripping system operated and the supply to the pressure cooker was automatically switched off again.

A further fault was introduced in this duplicate back-up system. The temperature pressure and voltage rose further until a *third* sensor operated (h), this time related to pressure rather than temperature.

So the apparatus has back-up on back-up, duplicate systems, diverse systems based on different principles. They all took care of our nuclear-heated pressure cooker.

The next step of the experiment was to cut out all the protection system and await events. In fact, the operator switched off the electricity supply. This illustrates the role of the operator. He monitors the plant and intervenes only if necessary.

Overall, this demonstration illustrated the Defence in Depth approach. Our pressure cooker was protected by a range of means, including the operator. And we still had the ultimate protection of the safety valve and plug ejection available.

Further protection comes from routeing the cables via different paths so that any fire or disruption does not itself disable the safety systems. This lesson was learned after the cable fire started by a candle at the Brown's Ferry nuclear plant in the US in the early 1970s.

So as a final part of the experiment, and to celebrate learning that lesson, fusible links in the protection system were melted by candle flame. The system tripped because, when operating normally, it is actually designed to fail-to-safety.

This all seems very elementary and simple and it is. And in just this way, Defence in Depth and fail-to-safety design are essential principles of nuclear reactor protection and the reactor operator is an integral part of that system.

The simulator (Demonstration 2)

To add weight to the last illustration, the equivalent of the pressure cooker demonstrations is possible on the reactor simulator.

Let me remind you first that the indications received on the simulator are those which the AGR reactor operator would receive. Up to a point that is, because in the next experiment I deliberately prevent the shutdown system operating. Disabling the shutdown system permits me to show the reactor trip signals operating one after the other as successive safety limits are approached and signalled. However, I must emphasize that, because of the very high reliability of reactor shutdown systems,

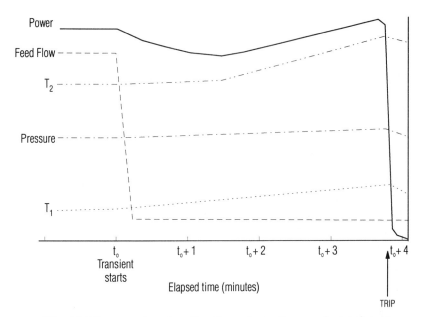

Fig. 22 Diagram showing the time dependence of simulator parameters.

this would be a very, very unlikely event. Since their inception British power reactors have never failed to shut down on demand.

Figure 22 shows the time dependences of reactor power, boiler feed flow, reactor gas inlet temperature (T_1), reactor gas outlet temperature (T_2), and reactor gas pressure.

After a period of steady state full power operation, I introduced a very severe fault, 80 per cent reduction in boiler water feed. Essentially this cut off the prime means of cooling the reactor. As the simulation progressed, warning lights appeared on the simulator fascia (Figure 23), indicating that a reactor trip was imminent.

Turning back to Figure 22, as boiler feed was reduced, reactor temperatures T_1 and T_2 and therefore the fuel temperature increased, giving the self-regulation power reduction discussed earlier. Later, moderator temperature increases actually reversed this trend. But this did not cancel the self-regulation because the graphite temperature changes were quite slow.

Allowing the simulation to run on showed the Reactor Protection System at work in sequence and warning lights appeared on the fascia (Figure 23).

> Reactor gas inlet temperature rise, trip imminent.
> Reactor gas outlet temperature rise, trip imminent.
> Reactor gas inlet temperature rise, trip occurred.

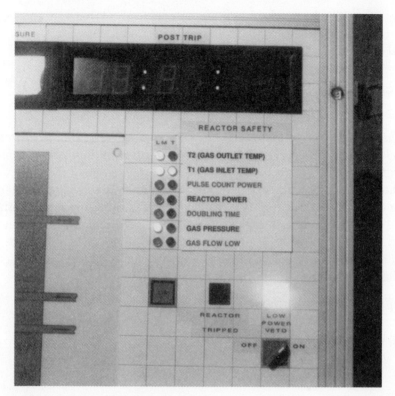

Fig. 23 Fascia indicator lights.

Finally, in recognition that the protection system was not operating satisfactorily, the reactor was shut down manually using the 'trip' control.

Of course, a real AGR suffering such a severe loss of feed would shut down in seconds (that is, the control rods would enter the core) as soon as the first trip level had been reached. The demonstration is intended to show the many other diverse ways in which the fault would be detected and the reactor tripped. In fact, AGRs have trip sensors directly monitoring boiler feed flow which were not modelled on the small simulator we used. If they had been, the demonstration would have been over in seconds and very uninteresting!

A fault such as this would generate a significant number of audible and visible alarms. The rate of feed flow to the boilers is a critical parameter which is clearly displayed to the operator. In practice he would act promptly in response to this. I will say more about the role of the operator in a moment, but I want to make one point crystal clear at this stage: for loss of cooling faults such as these, no claims whatsoever are made on the operator for the detection of the fault and the safe shutdown of the reactor. This is achieved entirely automatically. What was illustrated was the extensive nature of the back-up protection, Defence in Depth, provided by the diverse systems available in succession to trip the reactor.

The operator

(*In this chapter the pronoun 'he' has been used to describe the operator. This is for convenience only and does not imply that there are no female operators.*)

The operator is a further essential ingredient in the safety system's Defence in Depth. His role is a management task of information gathering, planning and decision making, and occasionally more active control when routine operation is disrupted. Operators are highly trained, on simulators and on actual plant, and are regularly tested for competence.

Their duties can be described as follows:

> monitoring normal operation to ensure plant is running within the conditions laid down and no abnormality is developing;
>
> manoeuvring the plant according to defined procedures and within approved limits;
>
> in abnormal circumstances monitoring the actions of automatic control and cooling systems and plant responses;
>
> intervening if action of the automatic systems is inadequate.

It is a requirement in the design of the most recent British reactors that the operator should not need to intervene to control an abnormal condition for a period of 30 minutes after it begins. The automatic systems are designed to achieve this. During this period the operator needs essentially to monitor the proper functioning of the safety systems. He takes action only if the response of these systems is judged inadequate for some reason.

We regularly test the operators' ability to deal with the unexpected, sometimes simulating very extreme circumstances.

The fault sequence in the simulator demonstration is one such sequence. The protection shutdown systems were not in operation. Reactor gas temperatures and reactor power continued to rise; audible and visual alarms were signalled. A well-trained operator would not simply have ignored these signals. He would have recognized that the automatic reactor protection system had not operated as it should and tripped the reactor promptly himself.

So tripping the simulator manually in the demonstration illustrated one obvious point, the role of the operator as part of the multiple back-up safety system. It also illustrated another, less obvious point. Too often we see the operator as a potential source of error. This is not surprising as the action of the operator is always questioned after any accident, nuclear or not. But the intervention to trip the reactor emphasized the valuable and positive role of the operator; his ability, unrivalled by any automatic system, to diagnose a new problem and to take appropriate action.

Conclusions

What conclusions can we draw?

In this discourse I described the fundamentals of nuclear reactor safety as embodied in the three basic principles:

> CONTROL the reaction,
> COOL the fuel,
> CONTAIN the radioactivity.

I demonstrated a selection of means by which these are implemented using simple principles and well understood phenomena and automatic safety systems complemented by operator action. I emphasized especially the means of achieving very high reliability.

The UK has been successful in the application of these principles. The unblemished safety record of our nuclear power generating plant over the past 30 years speaks for itself. So where do we stand in relation to my question 'What's the problem?' From my position I would say the potential enemy is complacency. It is important to guard against complacency and to build our defence to prevent complacency.

The nuclear industry has learned much from the major accidents at Three Mile Island and Chernobyl. Following Three Mile Island, the Institute of Nuclear Power Operations (INPO) was set up in the US with the express objective of achieving the highest standards of safety in US plants. Many overseas utilities, including Nuclear Electric, are associate members of this Institute.

Following Chernobyl, this concept was extended on a global basis by the setting up of the World Association of Nuclear Operators (WANO), to which virtually every utility operating nuclear plant belongs. Its mission is continuous ongoing improvement of nuclear safety worldwide, brought about by the sharing of experience and the process of bench-marking and peer review between operators.

In the UK, we participate fully in these international initiatives to improve nuclear safety. Experience on events with safety significance is freely exchanged between stations and operators. This approach is mirrored in our UK practices. Operational experience is analysed, lessons are learned and fed back. Best practices are shared between companies and stations. Detailed examinations and evaluations of plants are routine, carried out by fellow operators in a peer review process involving scrutiny of station practices. All these activities are antidotes to complacency.

So far I have not mentioned the role of the UK industry's safety watchdog—the Nuclear Installations Inspectorate (NII). In the UK, as with most other countries, operation of a nuclear installation is a licensed activity. It is not the role of the NII to try and double-guess the designers and operators of the plant. Their main task is to see that the operator has the means, the resources and the policies in place to ensure safe operation and that he does indeed carry out these policies in an effective manner. Sometimes, rather unfairly perhaps, the NII have been likened to 'a nagging wife'. Often their attentions are irksome. But they are always highly expert and are a major antidote to complacency.

Another driving force for safety is our recognition in Nuclear Electric that business success is directly linked with safe operation. In the nuclear power industry we have long recognized that if we are not safe we have no business and no future.

JOHN COLLIER

Born 1935, educated at St Paul's School, Hammersmith. Spent the first 30 years of his working life primarily with the United Kingdom Atomic Energy Authority, rising from engineering apprentice to Director of Safety and Reliability. 1983–6 was Director-General of the Generation Development and Construction Division of the Central Electricity Generating Board. Returned to UKAEA as Deputy Chairman, becoming Chairman in 1987. Appointed Chairman and Chief Executive of Nuclear Electric plc in 1990, and Chairman in 1992. He is a Fellow of the Royal Society, a Fellow of the Royal Academy of Engineering and a Fellow of the Institutions of Chemical, Mechanical and Nuclear Engineering. Holds an honorary Doctorate of Science from Cranfield Institute of Technology and an honorary Doctorate of Engineering from Bristol University. He is the 1993 Calvin Rice Lecturer and an honorary lifetime member of the American Society of Mechanical Engineers.*

* John Collier died from cancer on 18 November 1995, aged 60.

Modelling reality with supercomputers

C. RICHARD A. CATLOW

The images shown in Plate 6 are models of star formation. They simulate the condensation of an enormous gas cloud into a cluster of stars which then dance around each other under the attraction of their mutual gravitational fields. The model was generated by the Cardiff astrophysics group [1] using a supercomputer at the Atlas Division of the Rutherford Laboratory. It has compressed this extraordinary event, whose extent in both in time and space is almost unimaginable, to a level which we can understand. But in understanding the complex world in which we live, mankind has always needed models. Models allow our minds to explore the reality that they represent; and they may scale huge objects such as stars and tiny objects such as atoms to a size which we can comprehend and become familiar with. Models are therefore essential in the scientific quest for a rational understanding of the physical universe. Visualization is, moreover, an immense aid to the scientific imagination, as it seeks new relationships and connections between concepts and phenomena— the process which is at the heart of scientific discovery.

Models have been used since the beginning of scientific thought, and some of the earliest and most detailed scientific models relate to astronomical observations. The earth centred cosmologies of the ancient world were perfected in the Ptolemaic system of the universe shown in Fig. 1. This crude representation of a remarkable and complex structure model explains many observations on the motion of planets and stars. Although one has some sympathy with King Alfonse of Castille who remarked after receiving an explanation of the system, 'If the almighty had consulted me before embarking on the creation, I would have recommended something simpler.' The King was of course right, because the Ptolemaic system was swept away by the Copernican revolution in the sixteenth century, which produced far simpler heliocentric models for

Fig. 1 Model of the Ptolemaic system of the Universe.

the solar system which of course explain astronomical data more straightforwardly and could subsequently be rationalized by Newton's gravitational theory.

In astronomy and cosmology, which are amongst the most enduring scientific grand challenges, models are therefore crucial. 'Global modelling' is also playing an increasingly important role in newer sciences which aim to understand how the earth's atmosphere, oceans and interior work. For example, Plate 7 shows a beautiful example of a computer generated model of the circulation of the oceans around the Antarctic continent [2] and the key role of computer modelling in predicting the behaviour of the atmosphere is well known. Their role in engineering and applied sciences is obvious, but contemporary technology is expanding enormously their range and sophistication, as in, for example, the model of the distribution of pressure over the wings of a fighter aircraft in flight as shown in Plate 8; such models are, of course, an essential design tool in modern aeronautical design.

These are exciting and rapidly developing fields; but the present article will give pride of place to my own research interest in understanding matter at the microscopic level—the world of atoms and molecules—and in revealing the marvellously varied ways in which atoms combine to give structures of immense complexity and beauty; structures which support life like the enzyme lysozyme, shown in Plate 9, whose atomic architecture was first elucidated at The Royal Institution in 1965; and structures which lead to extraordinary and technologically important properties, for example the oxide material lanthanum copper oxide whose crystal structure is shown in Plate 10 and which, when doped with strontium, shows the extraordinary phenomenon of super-conductivity—conduction of electricity without resistance—at temperatures of approximately 40 degrees above absolute zero, far higher than had previously been achieved with complex metallic alloys. The discovery of the material by Bednorz and Muller [3] was one of the major breakthroughs in solid state science and initiated a fever of research which has led to materials showing superconductivity at temperatures as high as 160 K.

Our ability to model and explore complex three-dimensional structures like these has been revolutionized in recent years by new computer technology, which allows us to construct, display and manipulate models at a level of detail which was previously impossible. Indeed, computer modelling and graphics are influencing almost all aspects of science and technology. But the role of computers in contemporary science is more profound. The enormous achievements of science over the last few hundred years—many of which originate from The Royal Institution—have given us an increasingly accurate knowledge of the fundamental forces of nature. To achieve this, it has often been necessary to strip down nature to its essentials—to simplify reality. Given this fundamental understanding obtained by the strategy of simplification, we can now move to an even more exciting phase of exploring the complexity which we see in the world around us. And this is where computers play such a central rôle. The constantly expanding processing power and memory of modern computers allows us increasingly to translate fundamental knowledge into models of reality. Figure 2 attempts to schematize the approach. The input used in generating the model is the basic laws of physics and databases of relevant information on the system. The processing power of the computer allows us therefore to explore, develop and display a model specified by the scientist—a model that obeys the basic laws of science. Crucially, the model then confronts reality, and is refined and improved. A deeper and more predictive understanding of nature is developed.

The computational scientist is, of course, almost uniquely privileged in that the power of the technology which he uses continues to grow almost explosively. When I lectured at The Royal Institution four years ago [4], I pointed out that the supermini computer recently installed in the basement of The Institution had a power greater than that available to the whole of the University of London in the 1970s. The Silicon Graphics Challenge machine installed earlier this year will, following a recent upgrade, have a power ten times greater than that supermini. And

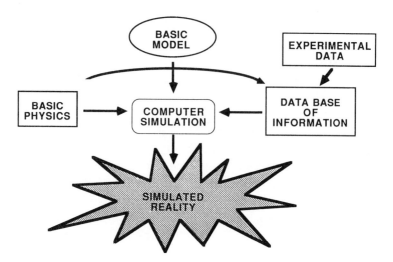

Fig. 2 Schematic representation of the development of scientific models using computers.

we will see in a few minutes some of the ways in which we are trying to exploit this marvellous facility. Does this expansion in computer power promise to continue indefinitely? It certainly will continue into the fore-seeable future. And one reason for this is that the computer industry is exploiting a simple but very effective idea—the idea summed up in the proverb 'Many hands make light work.' One (but only one) strategy in modern computer hardware is to use the principle of parallelism—to distribute a huge complicated task to a large number of modest (and cheap) processors. If the computational task is such that the processors can work independently for a period of time and if the different pro-cessors can communicate their results rapidly to each other, we can achieve high performance with only modest technology for each pro-cessor.

This vision of massively parallel processing has been achieved; and recently at Edinburgh the most advanced academic computer installation in Europe was opened—the CRAY T3D in which 512 powerful DEC processors collaborate and communicate to undertake enormous computational tasks, organized as shown in Plate 11.

This computer and the ones which follow it provide an immense opportunity for scientists in this country; an opportunity which, I think, is equivalent to that in 1808 when funds were raised by subscription for a powerful battery shown in Fig. 3 to be installed in The Royal Institu-tion, which allowed Davy, and subsequently Faraday, to develop their amazing discoveries in electrolysis and electromagnetism.

Now let us look at how the field is developing, the problems that it is

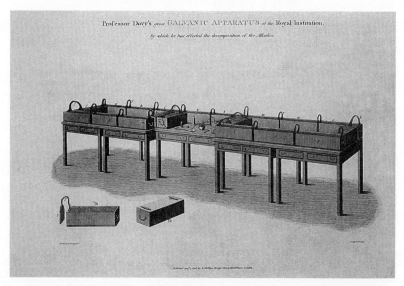

Fig. 3 Battery built in 1808 for Sir Humphry Davy using funds generated by public subscription.

tackling, and the future prospects. First we will consider 'atomistic' modelling, constructing models for the complex reality of matter at the atomic level. In this case the paradigm I have presented is perfect: we understand the fundamentals which control the behaviour of atoms and molecules. As discussed in my previous article in the Proceedings of this Institution [4], there are two approaches to modelling atomic assemblies. The first and more fundamental method is to solve the 'Schrödinger equation' to give the distribution and energies of electrons in the field of the atomic nuclei. The growth of computer power, coupled with the developments in algorithms and methods has led to rapid progress in the range and precision of such calculations. However, even with the biggest and most powerful computers, the requirements of such calculations still limit them to systems containing tens rather than thousands of atoms. But for larger systems we may adopt an alternative and highly effective strategy. We represent the interaction between atoms in terms of interatomic potentials which describe the dependence of the energy of the assembly of atoms on its geometry. Such potentials may be obtained either from more fundamental calculations (of the type described above) or from experimental data. And extensive databases of potential functions are now available. As illustrated in Fig. 4, using both approaches the computer can develop and display models for molecules and materials.

Our first need, if we are to understand matter at the atomic level, is to

MODELLING of MATTER at the ATOMIC LEVEL

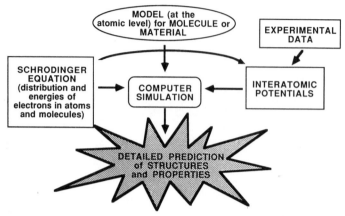

Fig. 4 Schematic representation of the development by computational techniques of atomistic models of molecules and materials.

know the structure, the arrangement of the atoms in the molecule or solid. I want first to look at crystalline solids where the arrangement of the atoms is regular and repeating. In my earlier article [4], I said that one of our ambitions in the modelling of solids was to be able to predict the structure at the atomic level of a crystal from knowledge only of its constituents. This field has progressed enormously since then; and a good recent example is provided by the work of Tim Bush at The Royal Institution. The real challenge in 'predicting' structures is the generation of approximate, plausible models which can be refined by the technique of 'energy minimization'—a standard computational procedure which locates the lowest energy configuration to the starting point of the calculations. Work of Freeman discussed previously [4] used 'simulated annealing' techniques to take an initial random distribution of atoms and explore ways of arranging the atoms in a manner which favours distributions with low energies. Bush [5] has used an alternative approach based on 'genetic algorithms' (or evolutionary programming techniques) which modify structures (again starting from a random starting point) in a systematic manner, allowing them to evolve towards plausible models, which are then subjected to energy refinement. The method was applied successfully to determine the structure of a complex oxide Li_3RuO_4, which had previously eluded solution. The resulting structure is shown in Plate 12. The application of computational techniques to developing models for structures of even greater complexity is likely to become increasingly important and routine in the future.

We have also made great strides in our ability to generate models of glassy materials. Here I will highlight the work published in 1992 by Behnam Vessal and co-workers [6] on silicate systems. They investigated sodium silicate glasses and used the computer to simulate the melting of the crystalline material and then simulated a rapid quench—the real way we make glasses. Plate 13 illustrates the resulting model of the glass. This beautiful if disordered structure shows fascinating features—the rubidium ions group together into channels—a process which had been proposed earlier by Neville Greaves and now emerges naturally from the computer models.

In atomistic modelling of *biological structures*, there has been equally impressive progress. Here are just two examples; we will see more later. The first is a snapshot from a very recent simulation by Tim Forrester on the CRAY T3D at Edinburgh of a cell membrane. Membranes are double layers of molecules with long tails. The tails point inward to escape from surrounding water; the ends point outward because they interact favourably with water. It is important to be able to move ions across the cell membrane; and the simulation, a snapshot of which is given in Plate 14, shows a molecule, valinomycin, which can capture potassium ions and transport them through the membrane.

Many other exciting examples are to be found in the field of molecular biology, for example, the simulation work by Julia Goodfellow at Birkbeck College of DNA—the most important molecule in the Universe as it is of course the basis of the genetic code, where a detailed modelling of the molecule shows how it flexes and bends [7], a vital process by which literally metres of the molecules are packed into micrometres in chromosomes.

Of course, what determines the behaviour of molecules and materials in many circumstances is not what happens in their interior, but what happens on their surface. *Chemical reactions* often take place on surfaces; and surfaces can promote chemical reactions, as in the field of heterogeneous catalysis which contributes so much to the chemical industry and hence to our lives. *Friction*, the overcoming of which consumes huge amounts of energy, takes place between surfaces sliding over each other. *The growth of crystals* takes place at surfaces; and if we want to prevent it, we need to modify the surface. And computers are guiding and illuminating our knowledge of surface structure and behaviour.

We will look first at our ability to model the structure and energies of the surfaces of crystals, which we will show leads into modelling their shape, or morphology. New computer codes developed at The Royal Institution by David Gay and Andrew Rohl [8] offer new opportunities for modelling complex systems. To illustrate current capabilities, we take a common and important mineral, quartz, found of course in nature,

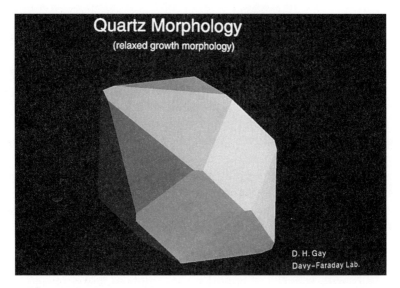

Fig. 5 Calculated morphology of α quartz.

but synthesized as an industrial material. Gay and Rohl have simulated the surface structure of quartz, shown in Plate 15. Interestingly, the surface requires hydroxylation (conversion of the surface oxygens to hydroxyl groups) to achieve stability. The figure also illustrates the substantial rearrangement or 'relaxation' of the atoms in the surface region.

Using the calculated surface energies, we are also able to calculate the crystal morphology or shape. Thus the predicted morphology for quartz is shown in Fig. 5. It is a remarkably faithful model of reality. Real quartz crystals have the same faces in the same proportion.

We mentioned above the importance of crystal growth and catalysis: both of these depend on molecules or atoms docking on to surfaces. We can show how far we have come in simulating these crucial processes by demonstrating recent work of Andrew Rohl in modelling the interaction of molecules with the surface of $BaSO_4$. Why are we interested in this system? Mainly, in the present context, because of one of its most basic chemical properties—its insolubility. Because it is so insoluble it precipitates, for example in oil pipes, and for obvious reasons this is a major problem in the oil industry. But we can prevent this precipitation by adding inhibitors. And recent work at The Royal Institution has shown how these inhibitors work—how they 'dock in' on the surface, as illustrated in Plate 16 for the case of the commonly used diphosphonate inhibitor on the surface of $BaSO_4$. Once such molecules have docked at the surface, they block further crystal growth. Of course the challenge is

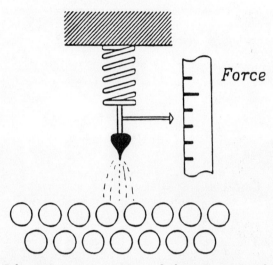

Force

Fig. 6 Schematic representation of the operation of probe microscopy techniques in which an atomically sharp tip scans over the surface at constant force.

to design new materials, new inhibitors, which block more effectively; and computer modelling techniques make this a real possibility.

These are computer models of surfaces; how do we learn about them experimentally? If you want to learn about the structure of an object, one strategy is to poke it with something sharp. This idea has been the basis of an increasingly widely used range of techniques for imaging surfaces, known as scanning probe microscopy, the basic idea of which is illustrated in Fig. 6. Here an atomically sharp tip scans over the surface. And in the version of the technique we will now consider, known as atomic force microscopy, as it scans the surface it is constrained to have a constant interaction force with the surface. To do this, the tip must move up and down; and these motions are in response to the structure of the surface, and can be translated into an image of the surface [9] as shown in Fig. 7. But if we are to interpret these images, we need to be able to model the interaction between the surface and the tip. Recent work at The Royal Institution of Alexander Shluger, David Gay and Andrew Rohl [10] has shown how we can simulate tip–surface interactions. In their work a model of the tip of magnesium oxide scans over a rock salt (sodium chloride) surface maintaining a constant tip–surface force. A snapshot from the scan is shown in Plate 17. The simulation models the upwards and downwards motion of the tip to produce a simulated image. The simulations are of particular value when they model the scanning of the tip across surfaces with defects, impurities and steps, where they show that sometimes the tip can drastically

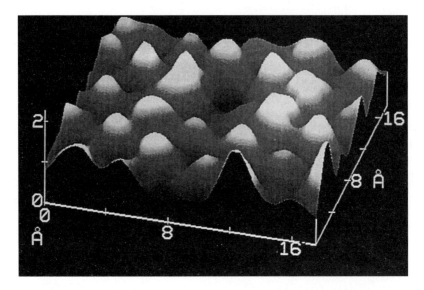

Fig. 7 Image of the surface structure of sodium chloride generated using atomic force microscopy (from reference 9).

perturb the surface as illustrated in Plate 18. Clearly this kind of information is vital if we are to plan and interpret these experiments properly.

The most important aspect of surface science is perhaps that chemistry can happen at surfaces. Indeed, the whole field of heterogeneous catalysis is about chemical reactions taking place at surfaces—processes which are at the heart of the chemical industry and which produce £400 billion of products worldwide each year. To promote the desired chemical reaction, to make it selective and specific for particular molecules, we often need to modify our surface, to engineer it at the atomic level. One example is the Ziegler Natta catalysts which convert the gas ethylene into a plastic—the polymer polyethylene. Here, methylated titanium chloride molecules are deposited onto a magnesium chloride surface. Computer simulations at The Royal Institution by J. S. Lin have shown how the titanium-containing molecules bind to the surface, creating sites for coordination of the ethylene molecules, as shown in Plate 19. We are currently investigating how additional ethylene molecules insert themselves, leading to a growing chain.

A second interesting strategy is to take one type of crystal and add a single layer of a second type onto its surface. In order to cohere, to stick to the crystal below, this layer has to change its structure significantly, and these changes may help to promote catalysis. We will consider one example, a single layer of vanadium pentoxide deposited onto titanium

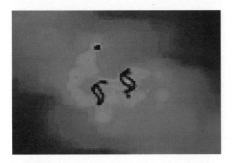

Plate 6 Computer simulation of the condensation of gas clouds to form a star cluster (after reference 1). The red regions represent high densities; the image shows two protostellar clusters in orbit around each other.

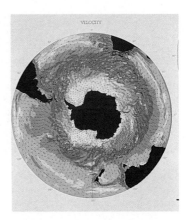

Plate 7 Model of the circulation of ocean currents around the Antarctic continent.

Plate 8 Computer model of the distribution of pressure over a fighter aircraft in flight. Blue colours indicate regions of low pressure; red colours show high pressure regions. (Kindly supplied by CRAY Research (UK) Ltd.)

Plate 9 The molecular structure of the enzyme lysozyme with a sugar molecule bound at the active site.

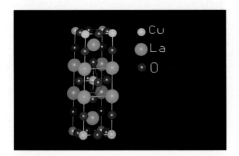

Plate 10 The crystal structure of the superconductor lanthanum copper oxide (La_2CuO_4); red spheres are oxygen atoms, blue spheres are lanthanum and yellow spheres are copper.

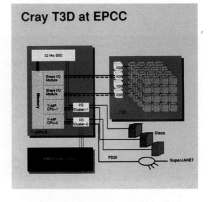

Plate 11 Schematic representation of the connectivity of the CRAY T3D massively parallel supercomputer. The right-hand side represents the interconnected processors.

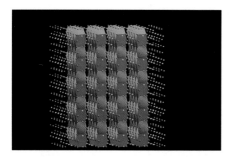

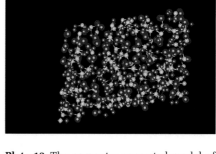

Plate 12 The crystal structure of Li$_3$RuO$_4$ generated from an initial random arrangement of atoms. The purple polyhedra represent RuO$_6$ octahedra and the turquoise polyhedra show the coordination spheres of the lithium ions.

Plate 13 The computer generated model of glassy Na$_2$Si$_2$O$_5$ (from Vessal *et al.* [6]); blue spheres are sodium ions, red spheres are oxygen and yellow spheres are silicon.

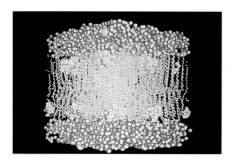

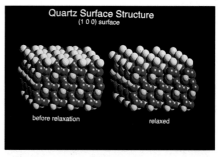

Plate 15 Simulated structure of the (100) surface of α quartz (SiO$_2$): red spheres represent oxygen atoms, white spheres hydrogen atoms and yellow spheres silicon atoms. The structure on the right was generated by allowing the atoms in the surface region to relax to their equilibrium configuration.

Plate 14 Snapshot of a simulated membrane containing diffusing molecules (kindly supplied by Tim Forrester and Julian Clarke).

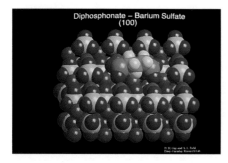

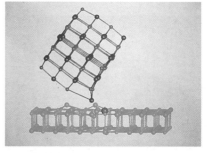

Plate 16 Growth inhibiting diphosphonate molecule bound to the surface of BaSO$_4$.

Plate 17 Snapshot from a simulated scan of a magnesium oxide tip scanning over the surface of sodium chloride. The tip is close to a surface impurity ion (shown in purple).

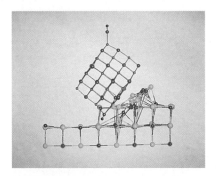

Plate 18 Simulated interaction of the tip with a step on the surface of sodium chloride.

Plate 19 Methylated titanium chloride on the surface of magnesium chloride interacting with an ethylene molecule (lower image). White spheres are surface chloride ions; the blue sphere is a titanium atom; yellow spheres are chlorine atoms bonded to titanium.

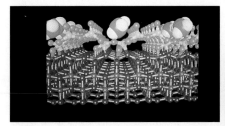

Plate 20 Simulated structure of supported vanadium pentoxide monolayer on titanium dioxide (anatase). Absorbed ethylene molecules are also shown.

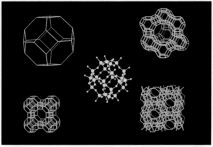

Plate 21 Crystal structures of some common zeolites. The sodalite cage is shown in the centre and the top left; zeolite A, bottom left; zeolite Y, top right; zeolite ZSM-5, bottom right.

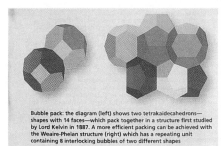

Bubble pack: the diagram (left) shows two tetrakaidecahedrons—shapes with 14 faces—which pack together in a structure first studied by Lord Kelvin in 1887. A more efficient packing can be achieved with the Weaire-Phelan structure (right) which has a repeating unit containing 8 interlocking bubbles of two different shapes

Plate 22 Structure of the most energy efficient bubbles in soap foams, after Dennis Weare and colleagues, Trinity College, Dublin.

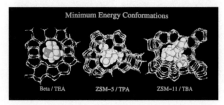

Minimum Energy Conformations

Beta / TEA   ZSM-5 / TPA   ZSM-11 / TBA

Plate 23 Templates docked in the pores of zeolitic solids. In each case we show the template which is used in synthesizing the particular zeolite.

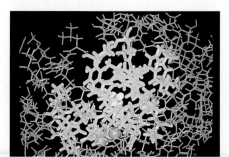

Plate 24 The enzyme methanemono-oxygenase containing a methane molecule within its groove. Pink spheres represent the iron active site. The carbon of the methane is green.

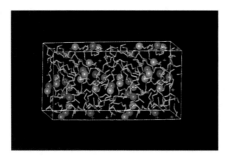

Plate 25 Sodium (blue) and iodide (pink) ions dispersed within a matrix of amorphous polyethylene oxide.

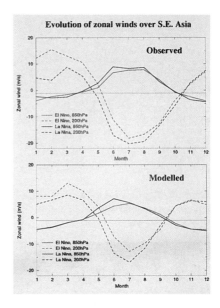

Plate 26 Computer model of wind patterns over South East Asia, generated by the UK UGAMP project. The observed and simulated annual variations are in good agreement (kindly supplied by Dr L. Steenman-Clark).

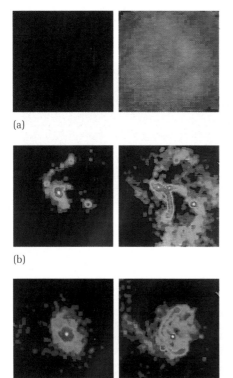

Plate 27 Simulation of the condensation of primeval gas cloud into a galaxy; the left hand images in each case show the density of stars, the right hand the density of gas. The three states shown are separated in time by approximately 1.5 billion years (kindly supplied by Prof A. Nelson).

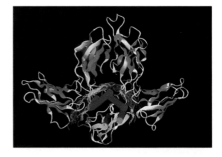

Plate 28 Ribbon diagram of the IL2 complex. The four long helices of IL2 are in red, while the short helix of the AB loop is magenta and the short β-strands pink. The α chain is yellow and white, the β chain blue and cyan, and the γ_c chain green. The surface of the area buried between IL2 and the β and γ_c chains is shown by white stippling. (Photograph courtesy of Dr Graham Richards.)

dioxide; it is widely used in the chemical industry to convert hydro-carbons such as ethylene to aldehydes. These systems have been studied recently by Dean Sayle at The Royal Institution. Plate 20 shows the computer generated model for this structure, onto which ethylene molecules have been adsorbed. The structure of the vanadium pentoxide mono-layer is considerably different from that of the unsupported surface of the material, allowing surface oxygens to approach more closely to the adsorbed molecules, hence promoting the catalytic reactions.

One of the most ambitious of recent calculations was undertaken by Mike Gillan, Mike Payne and co-workers [11] who looked at how the chlorine molecule dissociates on the surface of silicon. By using parallel computers they were able to follow the process in enormous detail to study the dynamics of this dissociation and to watch how the electrons are redistributed throughout the process. Calculations of this type offer the possibility of getting to the heart of reactions on surfaces.

Now let us move to crystals which are all surface—zeolites and other microporous solids. Examples of these beautiful crystal structures are illustrated in Plate 21. A number of the more important zeolite structures are constructed by bridging or fusing together sodalite cages, the structure of which is also illustrated.

The sodalite cage has an interesting history. In 1887 Lord Kelvin tried to solve an intriguing problem associated with the structure of foams. If we consider a foam made up of bubbles all of which have the same size, what shape gives the lowest energy, most stable, foam? Kelvin came up with this structure. In fact Dennis Weaire and his colleagues at Trinity College Dublin have recently shown that another type of shape, shown in Plate 22, just beats the sodalite cage; and interestingly, this cage is used by other crystal structures known as clathrasils.

Because of the porous structures, molecules can diffuse into zeolites and they are sorbed and diffuse at different rates, so these solids are widely used by industry in the field as separators. They also catalyse chemical reactions of the sorbed molecules; and the nature of these reactions is controlled by the structure of the surrounding crystal—the phenomenon of shape selective catalysis.

The really exciting problems in zeolite science all relate to the behaviour of molecules inside these intriguing porous architectures. And we can model the behaviour of molecules inside zeolites in great detail. We can model how molecules can diffuse inside the pores of these materials—a crucial process in both gas separation and catalysis, as discussed in my previous article in these Proceedings.

But we can also get to grips with an even more complex and subtle aspect of the chemistry of these materials; indeed, we are beginning to understand the factors controlling their synthesis. Zeolites are synthesized from gels to which are added 'template' molecules—organic bases which direct the synthesis towards specific architectures. Recent work here at The Royal Institution (in collaboration with BIOSYM/MSI Technologies in San Diego) has shown how the computer can pick the best template for a particular zeolite, by docking different templates into a given structure and finding out which has the lower energy—which fits the best. Examples of simulated host–template complexes are shown in Plate 23. This offers the real opportunity of designing molecules to provide specific new structures for use in catalysis and gas separation.

Docking of molecules is also central to key biological processes. A recent example where computational work has played an important role concerns the enzyme methane monooxygenase which has the ability to convert methane into methanol. This process would be of great potential economically if it could be realized on a large scale as there are large amounts of methane in remote parts of the world. Being gaseous it is difficult to transport; but if we can convert it to methanol which is a liquid, the transport problem is much more straightforward. There has been much excitement recently regarding the remarkable ability of enzymes to effect methane to methanol conversion. Plate 24 shows how a methane molecule fits into the groove in the enzyme methane mono-oxidase; the active site consists of a pair of iron atoms. The simulation work of Ashley George and Howard Dalton has modelled the docking and diffusion of the molecules in the enzyme. If we can elucidate the key features of the enzyme's structure which are responsible for its activity, we may be able to design new catalysts which effect this economically important reaction.

We will return now to the chemistry of materials, where we will highlight an unusual class of plastics which conduct electricity. Most plastics, for example polyethylene, are insulators. However, materials such as polyethylene oxide in which we have dissolved a salt (sodium iodide), conduct electricity quite well. The current is carried by mobile ions, which can migrate through the polymer matrix. If we can improve the performance of these electrolytes, they will have a major impact on battery technology and we will be able to improve the energy density in small batteries such as the lithium batteries manufactured on an increasingly large scale by SONY, although the crucial feature is the cathode material, which is based on a clever piece of solid state chemistry, discovered by John Goodenough's group in Oxford in collaboration with the Harwell Laboratories in 1979.

To return to the science of these materials, recent work of Glen Mills at Keele has modelled the behaviour of ions in polymer electrolytes.

Their complex distribution is illustrated in Plate 25. Of course there is nothing new in ionically conducting solids. And you will not be surprised to learn that this class of material was first discovered in The Royal Institution by Michael Faraday in the 1830s. He worked on lead fluoride; here we have an extract from his notebook of 1835:

> 2379. Now examined the habits of the fluoride of lead in the opened platina vessel. Found that, when cold, it did not conduct electricity as a current (the V. Electrometer being the test); but that, when heated, it acquired conducting power even whilst solid, so that sparks could be taken against it and much gas was evolved at the V. Electrometer. It then acted as the sulphuret of silver, for the current heated it more highly and then it conducted more powerfully. . . .

We now know that lead fluoride is a good ionic conductor because the fluoride ions are mobile, and the way they move has been simulated using computer modelling techniques.

In the remainder of this article, I want to look at models of some rather large objects. Molecules are of the order of a billionth of a metre. But as I said at the beginning of the article, one of the reasons for constructing models is to allow our minds to cope with objects whose dimensions are vastly different from those with which we are familiar. First we will consider global modelling—modelling of phenomena of a planetary scale. (It is worth noting before we do so that atomistic modelling is contributing enormously to our understanding of the interior of the earth, especially the minerals from which the earth mantle is constructed, for example $MgSiO_3$ which comprises 40 per cent of the earth's volume and whose structure and properties have been successfully modelled by David Price's group at University College London.)

We discussed earlier the modelling of oceans. There is perhaps an even more powerful incentive for modelling our atmosphere. The quantitation accuracy of such simulations is also emphasized by recent work of the UGAMP project [12] modelling wind patterns over SE Asia, shown in Plate 26.

Finally we will look at modelling the largest of objects—galaxies. In Plate 27 we show how the Cardiff cosmology modelling group [13] has simulated the condensation of a primeval gas cloud into a galaxy, where spiral arms are clearly visible.

Let us now briefly come back to Earth. Our Founder, Count Rumford urged us to devote our efforts to inventions that are of use 'for the common purposes of mankind'. Here are two examples. The first is the work of David Gosman's [14] group at Imperial College who have done computational fluid dynamical simulation on flow processes in engines.

The engine to which the modelling studies were applied were suffering catastrophic failures of the gaskets. The simulations were able to show that this was due to development of hot spots, which then allowed successful redesign of the engine.

Our last example takes us back to the microscopic world. It is work of Graham Richards and coworkers at the University of Oxford [15]. They have studied the interaction of a hormone with a receptor site with a protein interleukin 4 (shown in Plate 28) which plays an important role in the regulation of the immune system. The type of detailed under-standing of biochemical processes at the atomic level so generated is of enormous value in the design and optimization of pharmaceuticals which mimic the docking process.

To conclude, let us recall a remarkably prophetic remark of John Tyndall in his address to the British Association in 1870.

> An intellect the same in kind as our own would, *if only suffi-ciently expanded*, be able to follow the whole process from beginning to end. It would see every molecule placed in its position by the specific attractions and repulsions exerted between it and other molecules, the whole process and its con-summation being an instance of the play of molecular force.

With modern supercomputing, we are beginning to realize Tyndall's vision.

Acknowledgements

I am most grateful to Robert Bell, Tim Bush, Ashley George, Clive Freeman, David Gay, Dewi Lewis, J. S. Lin, Andrew Rohl, Dean Sayle and Alexander Shluger for undertaking the science and providing the computer graphics used in the article. The graphics were prepared using the INSIGHT II program of BIOSYM/MSI.

References

1. Chapman, S., Pongracic, H., Disney, M., Nelson, A., Turner, J., and Whit-worth, A. (1992). *Nature*, **359**, 207–10.
2. Fine Resolution Antarctic Modelling Project, National Environmental Research Council (1992).
3. Bednorz, J. G. and Muller, K. A. (1986). *Z. Phys. B, Condens. Matter*, **64**, 189.
4. Catlow, C. R. A. and Freeman, C. M. (1991). *Proc. R. Inst. GB*, **63**, 51–71.
5. Bush, T. S., Battle, P. D., and Catlow, C. R. A. (1995) *J. Mater. Chem.*, **5**, 1269.

6. Vessal, B., Greaves, G. N., Marten, P. T., Chadwick, A. V., Mole, R., and Houde-Walter, S. (1992). *Nature*, **356**, 504.
7. Goodfellow, J. and de Souza, N., to be published.
8. Gay, D. H. and Rohl, A. L. (1995). *J. Chem. Soc., Faraday Trans.*, **91**(5), 925–36.
9. Durig, U., Züger, O., and Stalder, A. (1992). *J. Appl. Phys.*, **72**, 1778. Meyer, E. (1992). *Surf. Sci.*, **41**, 3.
10. Shluger, A. L., Rohl, A. L., Gay, D. H., and Williams, R. T. (1994). *J.Phys.: Condens. Matter*, **6**, 1825, Shluger, A. L., Rohl, A. L., Gay, D. H., and Williams, R. T. *J. Vac. Sci. Technol. B*, in press.
11. De Vita, A., Stich, I., Gillan, M. J., Payne, M. C., and Clarke, L. J. (1993). *Phys. Rev. Lett.*, **71**, 1276.
12. UK Universities Global Atmospheric Modelling Project.
13. Davies, J. R. and Nelson, A. H. (1993). In *Star Formation, Galaxies and the Interstellar Medium* (eds J. Franco, F. Ferrini, and G. Tenorio-Tagle), p. 353. Cambridge University Press.
14. Gosman, A. D., Kralj, C., Marooney, C. J., and Theodossopoulos, P. (1992). *MechE. Combustion Engines*, **C448**, 035.
15. Bamborough, P., Hedgecock, C. J. R., and Richards, W. G. (1994). *Structure*, **2**, 839–51.

C. RICHARD CATLOW

Born 1947, educated at Clitheroe Grammar School and Oxford (M.A. Chemistry, 1970, D.Phil. 1974), following which he became University Lecturer in Chemistry, University College, London (1976–85) and subsequently Professor of Physical Chemistry, a joint appointment between University of Keele and Daresbury Laboratory (1985–9). He was appointed Wolfson Professor of Natural Philosophy in the Royal Institution in 1989. He is a Fellow of the Royal Society of Chemistry, and received the 1992 Royal Society of Chemistry award in Solid State Chemistry. His research interests are centred around the investigation of complex materials using a combination of computer modelling, quantum mechanical, synchrotron radiation and neutron scattering techniques.

Galvani, Frankenstein and synchrotron radiation

J. BORDAS

What connects an eighteenth century medical doctor and scientist, the fictional creator of a humanoid creature and the light emitted by particle accelerators? The common link between Luigi Galvani, Victor Franken-stein and synchrotron radiation is the research carried out on muscle tissues, these versatile (Inset 1) biological motors whose function is to

---

Available now: Linear Motor

Rugged and dependable: design optimized by world-wide field testing over an extended period. All models offer the economy of 'fuel-cell' type energy conversion and will run on a wide range of commonly available fuels. Low stand-by power, but can be switched within msecs to as much as 1 kW mech./kg (peak, dry). Modular construction, and wide range of available subunits, permit tailor made solutions to otherwise intractable mechanical problems.

Inset 1 Text of an advertisement to publicize a lecture given by Professor D. Wilkie to the Institute of Electrical Engineers some 30 years ago. This advertisement neatly summarizes the main properties of muscle tissues.

convert chemical energy into force and motion. In unravelling the con-nections between the three elements in the title, the aim is neither to describe accurately the current state of play in muscle research, interest-ing and exciting though this is, nor to address the many subtleties in the story of Frankenstein, nor to go in any depth about the production of synchrotron radiation (SR) but, rather, to illustrate how research can have unpredictable consequences.

Frog legs, sea water, copper and the navies

Luigi Galvani was born and worked in Bologna, where he became the Director of the Anatomical Museum (1768), a Research Associate in Anatomy at the University (1775) and Professor of Obstetrics at the Instituto delle Scienze (1782). After anatomical studies of, among other things, birds, kidneys and ears, Galvani devoted most of his time to physiological studies of muscles and nerves. From 1780 he investigated how these were affected by electricity and used for this purpose frog leg muscles. The muscles were prepared so that the inner nerve was partially exposed and when the nerve was grounded, electrical discharges caused muscular contraction. Galvani interpreted this discovery as proof for the existence of internal electricity in animals. Later on he found that contractions also occurred when muscle and nerve were connected via a bow made of two pieces of metal. At that instant he had discovered galvanism and simultaneously—in the form of a frog leg—the first galvanoscope, but for Galvani himself this discovery only confirmed the existence of animal electricity. When he described his discovery in *De viribus electricitatis in moto musculari commentarius*, 1791, Galvani equated the muscle to a charged Leyden jar and the nerve to its internal coating, and he claimed that a discharge takes place when the muscle and the nerve are joined together by a piece of metal. He assumed that different fibres would have opposite charges and that when discharged contraction would follow.

Galvani's explanation of muscular contraction based on the idea that some kind of animal electricity was responsible for the characteristics of being 'alive', soon captured the imagination of many individuals. Among these one finds the poet Percy Bysshe Shelley, who went to rather extreme forms of behaviour in order to put to good use the 'vital electrical fluids' which Galvani had postulated (Inset 2).

However, the Italian physicist Alessandro Volta (1745–1827) challenged Galvani's explanation. The basis of Volta's challenge was that muscular contraction occurred when two points on the nerve were joined together by two dissimilar metals, which led him to argue that it was at the contact between the two metals that the electrical currents originated. To prove his point Volta had to amplify the electrical currents to a measurable level and to do so he connected a series of zinc and copper plates and interspersed them with flannel disks soaked in brine. In doing this he had invented the voltaic pile and its successor the battery.

The great chemist Sir Humphry Davy (1778–1829) recognized the potential of the galvanic battery for his studies on chemical decomposition and, among other things, used this invention to develop the process

---

Choice of two control systems:

1) Externally triggered mode
Versatile, general-purpose units. Digitally controlled by picojoule pulses. Despite low input energy level, very high signal-to-noise ratio. Energy amplification 10^6 approximately. Mechanical characteristics (1 cm modules): maximal speed optional between 0.1 and 100 mm/sec; stress generated 2 to 5×10^{-5} Newtons m^{-2}.
2) Autonomous mode with integral oscillators
  Especially suited for pumping applications. Models available with frequency and mechanical impedance appropriate for:
a) Solids and slurries (0.01–1.0 Hz)
b) Liquids (0.5–5 Hz): lifetime 2.6×10^9 operations (typ.), 3.6×10^9 (max.), independent of frequency.
c) Gases (50–1000 Hz)
Many optional extras, e.g. built-in servo (length and velocity) where fine control is required. Direct piping of oxygen. Thermal generation. Etc.

Inset 2 Shelley influenced by the 'miracles' of galvanism experimented with the effects of electricity on his own body.

of galvanization. Sir Humphry Davy (*Philos. Trans.*, 1825) argued that 'Cu at sea oxidises because of the O_2 diluted in sea water, Cu oxide is formed. The oxide takes muriatic acid from the soda and magnesia in the sea water and it forms submuriates of the Cu oxide. Make Cu electronegative and it will not happen, therefore bring Cu into contact with Zn or Fe. The former becomes negative and the latter positive. A piece of Zn the size of a pea should be sufficient to protect fifty square inches of Cu. But Cu must not be rendered too electronegative, otherwise substances like Mg and Lime separate from the water and form a nidus for seaweeds and shell-fish.' This protecting process was applied to ships in service, resulting in very substantial savings for the navies. This chain of events, which started with the 'violent convulsions' of a frog's leg and ended with the invention of galvanization, illustrates the difficulties faced by policy makers in deciding a priori the potential for wealth creation of research driven by curiosity.

Animal electricity, life creation, Prometheus and the entertainment industry

One of the most unexpected results of Galvani's research is the novel *Frankenstein or the Modern Prometheus*. The start of the connection has to be traced to an ex-Dissenting minister turned atheist, William Godwin (1756–1836). Godwin had leapt from obscurity to fame with a celebrated attack on institutions contained in his work *Enquiry Concerning Political Justice* (1793). Although the establishment of the time regarded Godwin as just an unsavoury radical of the anarchist left, his ideas were tremendously influential with those intellectuals who were striving to find answers to the many social problems of the time. His influence was so notable that for a while 'wherever liberty, truth or justice was the theme, his name was not far off' (William Hazlitt).

Godwin's views on marriage were that it was 'a system of fraud' which sustained 'the most odious selfishness'. However, when his liaison with Mary Wollstonecraft, a pioneer feminist writer who had made her name with *A Vindication of the Rights of Women* (1792) and was admired by the radical intelligentsia of the day, resulted in her pregnancy, Godwin decided to marry. He argued, somewhat inconsistently, that he wished to protect her happiness 'which I have no right to injure' as otherwise nothing 'could have induced me to submit to an institution which I wish to see abolished'. Five months later (August 1797) a daughter, Mary, was born and Mary Wollstonecraft died ten days later of puerperal fevers.

Mary Godwin grew up in a highly stimulating intellectual environment which in many respects compensated for a fair amount of emotional deprivation. Sir Humphry Davy, in common with many other intellectuals of the time, used to visit William Godwin's household in Skinner Street, Holborn. Mary knew Davy's work well and through him the work of Galvani and Volta (for instance, it was Davy's *Elements of Chemical Philosophy* and *A Discourse, Introductory to a Course of Lectures on Chemistry* that she read in Autumn of 1816, close to the time when she conceived the story of Frankenstein). Other visitors included the recently married, hot-headed, eloquent, young man of nineteen, Percy Bysshe Shelley, whose hatred of tyranny and idealistic political notions drew him like a magnet to the aging radical philosopher's house.

The impact that Mary and Percy had on each other must have been instantaneous, because they first had occasion to socialize in May 1814; by June, Percy was almost a daily visitor to Godwin's home, with Harriet his wife virtually abandoned, and by July he and Mary had eloped to the continent. They returned to England at the end of the summer and by February 1815, Mary (aged seventeen) gave premature birth to their first child who died a few days later. In May 1816 Mary and Percy set out for their second continental trip, and settled in a cottage in Geneva. This cottage was adjacent to the residence of Lord Byron who had left England for good. A group consisting of Mary, Percy, Byron, Claire (Mary's step sister) and Polidori (Byron's personal physician) was formed and as the weather was foul, they had to spend significant periods of time confined in Byron's villa.

It was during one of these meetings that all the elements came together into the mind that was to produce Frankenstein. This was after Byron announced 'we shall each write a ghost story' and all agreed. Inspiration did not come easily to Mary, but eventually after being present at a conversation between Byron and Shelley where the possibility of bringing life back into a body was discussed, with the help of a nightmare and her background knowledge of Galvani's experiment (Inset 3),

---

Good to Eat
At Oxford . . .
'. . . Shelley proceeded, with much eagerness and enthusiasm, to show me the various instruments, specially the electrical apparatus; turning round the handle very rapidly, so that the fierce, crackling sparks flew forth; and presently, standing on the stool with glass feet, he begged me to work the machine until he was filled with the fluid, so that his long, wild locks bristled and stood on end. Afterwards he charged a powerful battery of several large jars; labouring with vast energy and discoursing with increasing vehemence of the marvellous powers of electricity, of thunder and lightning; describing an electrical kite that he had made at home, and projecting another and an enormous one, or rather a combination of many kites, that would draw down from the sky an immense volume of electricity, the whole ammunition of a mighty thunderstorm; and this being directed to some point would there produce the most stupendous results . . .'
Thomas Jefferson Hogg
The Life of Shelley (London, 1858), p. 33

Inset 3 Mary Shelley acknowledged the inspiration derived from Galvani's experiments.

she conceived the novel about the scientist Frankenstein. In this novel Mary Godwin emphasized Ovid's version of the Promethean legend, in which Prometheus is portrayed as a manipulator of mankind. In this approach she took issue, consciously or not, with the views of Shelley, her father and possibly with those of Sir Humphry Davy, as they all believed that Science was there to 'modify the beings' surrounding man. In fact, the character and words of Davy are replicated in the novel in the persona of Professor M. Waldman, Frankenstein's mentor and the man ultimately responsible for setting him forth on the path of his monstrous creation. In view of the usage that has been and, in all likelihood, will be made of some of the fruits of research, the ethical issues raised by Mary Godwin in her novel should be worthy of consideration by any scientist practising today.

Even though some respected writers (e.g. Sir Walter Scott) received the novel with enthusiasm, most of the ideas contained in it were, at the time, lost on the majority of the critics (Inset 4). Nevertheless, the story

---

Mary Shelley (Wollstonecraft Godwin) 30 August 1797–1 February 1851
'. . . Many and long were the conversations between Lord Byron and Shelley, to which I was a devout but nearly silent listener. During one of these, various philosophical doctrines were discussed, and among others the nature of the principle of life, and whether there was any probability of its ever being discovered and communicated . . .
. . . Not thus, after all, would life be given. Perhaps a corpse would be re-animated; *galvanism* had given token of such things . . .'
In the author's introduction to the standard novels edition in 1831.
Novels first published in 1818.
'. . . inculcates no lesson of conduct, manners, or morality . . .
. . . it fatigues the feelings without interesting the understanding; it gratu-itously harasses the heart, and only adds to the store, already too great, of painful sensation . . .
. . . the reader, after a struggle between laughter and loathing, [is left] in doubt whether the head or the heart of the author be the most diseased . . .'
The Tory Quarterly Review

Inset 4 An example of the kind of reception that Frankenstein received from some literary critics of the time.

became a bestseller overnight and, to the delight of many generations of viewers, gave rise to countless dramatic representations with enormous benefits to, among others, the film industry. Who could have predicted that Galvani's experiments were to inspire one of the classical pieces of nineteenth century literature, which is still providing the entertainment industry with a regular revenue?

Muscle research and synchrotron radiation

The research started by Galvani, and continued until the present day by countless other scientists, has not yet led to satisfactory explanation of the functioning of muscles. Here is where SR comes into play. So, what is SR and what does it have to do with muscle research?

SR is the light produced when electrons or positrons are accelerated. SR has been in existence probably since the beginning of the Universe. For instance, it is known that the emission of radiation (e.g. X-rays) by celestial bodies, such as the Crab Nebula, is due to such effects. This type of radiation is now harnessed as a major research tool and there are central laboratories (Fig. 1) where it is produced by centripetally accelerating high energy electrons or positrons using a magnetic lattice. The particles are kept in a circular orbit for many hours and whilst circulating emanate SR. The energy lost in this way is given back to the particles in the form of an accelerating electric field.

The physical basis for the production of SR is fairly straightforward. Let us imagine moving particles (electrons or positrons) which are continuously accelerated by magnetic centripetal forces so that the particles are kept in a circular trajectory. A charged particle undergoing acceleration will emit radiation according to Maxwell's laws. If the particle were travelling at relatively slow speeds the observer would perceive this radiation as that of a classical dipole emission and as such it would be emitted over a broad range of angles (Fig. 2, top panel). Let us assume that the particle is travelling at speeds approaching that of light and that an observer is looking at its emission along a tangent to its orbit. Here relativistic effects become important. We know that if θ_0 is the angle of emission of a given radiation ray measured from the line between the source and the observer, and if the source moves at a velocity towards the observer, then θ will be perceived at a different angle related by

$$\theta = \theta_0 \sqrt{(v/c)^2}$$

where c is the speed of light. Clearly when v equals c this angle will be zero. In other words the emission of the electromagnetic radiation as seen by the observer will have perfect collimation (Fig. 2, bottom panel).

One would expect that the frequency of the radiation is equal to that of the revolution frequency of the particle around its orbit (i.e. infrared frequencies). Here the Doppler effect (Fig. 3) comes into play and, in line with the common experience of hearing the sound of a car horn shifting towards higher pitches when the car is accelerating towards us, the frequency of the emitted radiation is continuously extended into the region of X-ray frequencies. How far the emission will extend into the high energy region is determined by how fast the particles are circulating.

Another important point is that in practice the particles circulate around their orbit packed in bunches with dimensions of the order of a few tenths of millimetres. Therefore the effective size of the light source

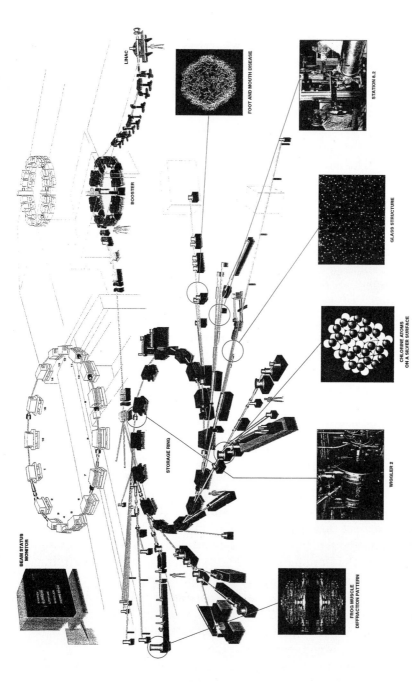

Fig. 1 An artist's impression of the Synchrotron Radiation Source (SRS) at Daresbury Laboratory. The electrons produced by an electron gun are injected into a booster synchrotron via a linear accelerator (Linac). The electrons acquire an energy of 600 MeV in the booster and are then injected into the main storage ring, where their energy is ramped to 2 GeV. The lattice of magnets keeps the electrons in a circular orbit. The radiation emitted from the tangent points at the bending magnets is used for experiments. The insets show some examples of the experiments and technology developments carried out at the SRS.

Synchrotron Radiation provides highly collimated electromagnetic radiation

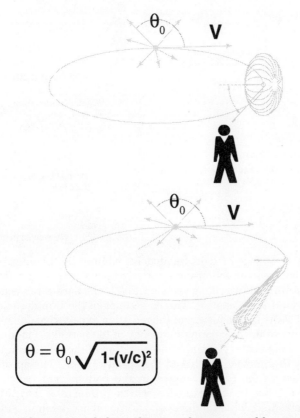

$$\theta = \theta_0 \sqrt{1-(v/c)^2}$$

Fig. 2 The top panel shows how an observer would experience the light emitted by the accelerated particles if their velocity v were smaller than that of the speed of light c. The bottom panel illustrates how the observer experiences the emission of light when the velocity of the light emitting particles approaches that of the speed of light. In this case the light emitted is greatly collimated because of relativistic effects.

is also that small. Moreover, in modern accelerators it is possible to have particle circulating currents of a few hundred milliampères.

Thus, the combination of the relativistic effects, the Doppler effect, the small source size and the high circulating currents lead to the production of a highly collimated, intense and continuous emission of light, which in practice extends from the far infrared to the hard X-ray region.

The brilliance of a source is defined as the number of photons it emits per unit time, per unit area and per unit solid angle in a certain

Synchrotron Radiation provides a continuous spectrum of electromagnetic radiation

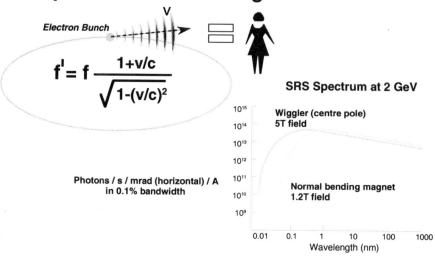

Fig. 3 The range of light frequencies emitted by the electrons undergoing acceleration continuously spans a spectrum from the far infrared (very long wavelengths) to the hard X-ray region (very short wavelengths). This is because of the Doppler effect which shifts the frequency of the radiation expected to coincide with that of the circulating frequency of the particles f to much higher frequencies f' owing to acceleration of the particles up to velocities approaching that of the speed of light. The spectral emission of the SRS is shown in the bottom-right panel. A wiggler is an insertion device which produces a very high magnetic field, which induces an additional acceleration on the circulating electrons. Note that this shifts the emission of radiation to even shorter wavelengths than that produced by the normal bending magnets.

wavelength band pass. One of the simplest consequences of Liouville's prin-ciple is that once an experimenter has defined what collimation, spectral resolution and beam dimensions are required for a measure-ment, the brilliance of the source predetermines how many photons are available for use. In other words, Liouville's principle shows that there is an upper limit to the number of useful photons delivered by a source and that this is determined by its brilliance. The brilliance of SR sources is several orders of magnitude higher than that of the most powerful X-ray generator and, consequently, the number of useful X-rays is at least 1000 times greater and in the most powerful accelerators now under con-struction or undergoing commissioning (e.g. the European Synchrotron

Radiation Facility at Grenoble) the intensities available for use are or will be many orders of magnitude higher.

So, what does all this have to do with muscles? Galvani's idea about animal electricity survived until the 1840s through Emile du Bois-Reymond's electrophysiological work and thereafter fell into disrepute. Nevertheless, the research that he started has remained an area of intense activity. Because of the complexity and sophistication of muscle it has proved an area of research in which progress has been very slow and always limited by the technical demands that it imposes on the experimenter.

Even though there is no general agreement about how muscle works, an enormous body of knowledge has been assembled. For instance, the structural details of the major constituents of muscle, the so-called thin and thick filaments, have been worked out. It is known that these filaments contain the motor proteins which are the molecules actin and myosin. Monomers of actin and myosin assemble during cell differentiation into these filaments which have structural organizations very close to that of helices. The thin filaments are made up of filamentous actin, which is a double helix in which two strands of actin monomers intertwine so that the structure has an approximate helical repeat every approximately 73 nm. This arrangement produces a long helical groove in which there is another filamentous protein called tropomyosin. In addition, at every 14 actin monomers there are two molecules of a protein called troponin (Fig. 4). The thick filaments are primarily constituted by myosin molecules. These consist of a long α-helical tail with a two headed end, made up of the so-called myosin S1 fragment. The myosin molecules assemble so that their tails pack into a backbone structure whilst the myosin heads protrude from it and approximately describe a three-stranded helix. At a given level along the thick filament there are three pairs of heads protruding from the backbone ('a crown'). These are disposed so that each pair is at approximately 120° around the backbone from the other two pairs. At subsequent levels the crowns are rotated by 40°, so that a three-stranded helix is formed, where each strand has a helical repeat every nine levels. However, there is an axial repeat at every approximately 43.0 nm.

The thin filaments are organized so that they emanate from a plane of proteins called Z-planes with opposite polarities. Two Z-planes delimit the 'crystallographic' unit cell of muscle tissues which is called the sarcomere. The sarcomere is bisected by a plane of density from which the thick filaments emanate also with opposite polarities. Where the thick and the thin filaments overlap they produce a hexagonal packing. This arrangement defined by the sarcomere repeats indefinitely in the muscle

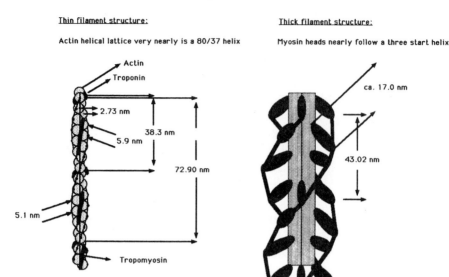

Fig. 4 A cartoon showing the helical structure of the thin (left) and thick (right) filaments of muscle. The two structures are drawn roughly to scale and approximate dimensions in nanometres are given.

tissue and this is why the sarcomere can be regarded as the crystallographic unit cell of muscle tissues. The classical work of A. F. Huxley and R. Niedergerke (1954) and H. E. Huxley and E. J. Hanson (1954) is to be thanked for the fact that we now know that muscle contracts by changing the degree of overlap between the thin and the thick filaments. However, even though we know that this is what happens, we do not yet know how the force necessary to achieve such sliding is generated. We know though that myosin works as an enzyme and that it hydrolyses the nucleotide adenosine triphosphate into adenosine diphostate and releases an inorganic phosphate in the process, and that the generation of force is associated with the dissociation of this nucleotide. But, how do these events cooperate so that the end result is the generation of force and/or motion?

There are many models of muscular contraction which have received varying degrees of recognition and acceptance, but the most robust of them all is that proposed by Sir Andrew Huxley in the 1950s and later refined and occasionally modified. The central elements of this model are that the relative sliding force between the thin and the thick filaments is generated by cyclic attachment between myosin heads and action molecules coupled with the hydrolysis of ATP and that, whilst this happens, the myosin heads behave as independent force generators.

A refinement to this model arises from the idea of the 'tilting head' developed by H. E. Huxley. In this model it is assumed that the force is produced by the active myosin heads undergoing some configurational change coupled to the hydrolysis of ATP which alters their attachment angle, thus producing a power stroke. In fact, it is known that the myosin heads, which are elongated objects 5.0–6.0 nm in width and 15.0–17.0 nm in length, are attached at an angle of approximately 45° in rigorized muscles. H. E. Huxley suggested that the configuration of the heads at the end of the power stroke corresponds to that seen in the 45° angled rigor state. Taking the length of the myosin head (subunit S1) to be about 15 nm, and assuming that during contraction the attachment angle changes from 90° to 45°, H. E. Huxley calculated that this would produce an axial translation of the head by about 12 nm for each molecule of ATP hydrolysed. This was in line with the value of about 16.0 nm estimated by A. F. Huxley to be the distance over which a crossbridge can generate force for sliding.

There is only one way that this mechanism of contraction can be tested on undisturbed muscle tissues and this is by using X-ray diffraction methods. However, muscle, like all biological material, is made of light elements and, because of this, the X-ray diffraction cross-sections are relatively weak. Also, and by common standards, it has an enormous crystallographic unit cell. This means that the diffraction features will be very closely spaced. Moreover, the contractile process occurs in a time scale of a few milliseconds and, therefore, to capture the desired information the data must be collected with time slices of about 1 ms. So, in order to carry out these measurements it is necessary to have an X-ray beam with high flux, high collimation and very small dimensions, in other words, very high brilliance. Because of the limitations in brilliance of conventional X-ray generators, this type of experiment would have remained a fantasy, but after about two decades of development of the required technology, it is now possible to carry out these measurements using the SR emitted by particle accelerators (Fig. 1).

It is interesting to note in passing that, keeping in with its history, muscle research has already had a beneficial impact in other disciplines. Muscle researchers were the first biologists/biophysicists to use SR and thus led the many other scientists who now use it for biological applications. Also, muscle researchers have been largely responsible for many of the technical developments now routinely used by other scientists. Pertinent examples are the techniques currently exploited in macromolecular crystallography and in the testing of industrial polymers.

So, how has SR helped our understanding of muscle function? SR has provided very compelling evidence for a series of molecular events associated with muscular contraction. Because the diffraction diagrams are collected as two-dimensional pictures in a movie-like fashion a lot of information about the molecular dynamics of the system has been accumulated. We know that there are at least four events associated with muscular contraction.

The first event takes place immediately following activation and involves the troponin and tropomyosin molecules moving relative to the position they occupied in the resting thin filament. The data suggest that signal transduction is initiated by troponin binding calcium which is released by electrical stimulation. This induces a movement of the troponin molecules which destabilizes the equilibrium in the system and as a result tropomyosin rolls away radially from its rest position. In doing so, it opens up the path for the interaction between the actin and myosin molecules. This process of thin filament activation occurs with a time constant of 14–16 ms.

Following thin filament activation, two almost simultaneous events take place with a time constant of approximately 30 ms, which includes a latency phase taken up by the activation events. These events are an order–disorder transition during which the register between the filaments is lost, and the formation of an acto-myosin complex. Finally the tension generating events set in. They occur with a time constant of approximately 52 ms and are characterized by a reorientation of the interacting myosin heads relative to the filament axis.

However, a striking feature of the diffraction patterns collected from muscle during contraction is the fact that the reflections at high angles are not only present during contraction but also that they become stronger (Fig. 5). This indicates that the myosin heads contributing to these reflections are aligned regularly with the actin monomers and implies that they must have a restricted range of axial displacement or tilt. Analysis of the data shows that the range of tilt cannot be greater than 5.0 nm. This effect cannot take place if a substantial fraction of the myosin heads interacting with actin and/or a major fraction of their mass is undergoing the kind of axial excursions demanded by the tilting head model. So, now one is faced with a dilemma. Do these results mean that a new hypothesis for muscular contraction has to be developed, or simply that the myosin heads that do the work are only a small fraction of the total and, consequently, they cannot be seen by X-ray diffraction. The latter possibility is somewhat disheartening as it is not possible to interpret what cannot be seen. The second possibility can be investigated and here SR will continue to play a major role.

Very recently, two major breakthroughs in crystallography have taken place and the atomic structures of the actin monomer and of the myosin

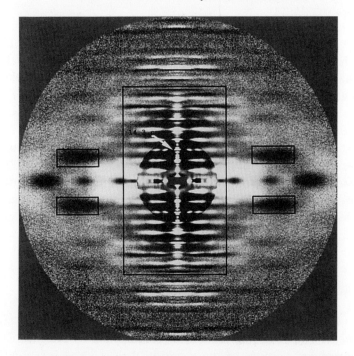

Fig. 5 Difference X-ray diffraction diagram from two time frames during the contractile cycle. This pattern corresponds to the subtraction of the data collected from muscles at rest and during the peak of isometric contraction. The lighter and darker areas correspond to the diffraction features disappearing and appearing during contraction respectively. The data provide a structural resolution extending from ca. 200.0 nm (given by the X-rays diffracted at very low angles; near the centre of the image) to ca. 2.5 nm (i.e. given by X-rays diffracted at the higher angles at the edge of the picture). Note that there are pronounced diffraction features appearing at high angles during contraction (darker areas at the edge of the diagram). The areas defined by the rectangles show the part of the diffraction diagram which can be recorded with time resolutions of a few milliseconds with current state of the art technology.

heads have been determined at atomic resolution (Kabsh *et al.* 1990; Rayment *et al.* 1993). This has opened up the possibility of carrying out some theoretical work that would have been otherwise impossible. In this context one should mention that by applying the Poisson–Boltzmann formulation it has been possible to calculate the distribution of electrical potentials on both the myosin heads and the actin filament. A somewhat surprising result has emerged, and that is that the actin filament is characterized by having a shroud of negative potentials, particularly prominent on the points where it interacts with the myosin heads, whilst the same type of calculations combined with molecular dynamics have shown that the myosin heads have pronounced positive potential bulges when they are in the ADP state. Rather conveniently, the calculations show that the electrical forces between the actin filament and a myosin head are of the order of a few piconewtons, which is very close to the experimentally determined force generated per head.

Whilst suggestive, the above cannot yet be regarded as an alternative hypothesis for contraction. However, it vindicates Galvani's view that muscle filaments are charged (even though he did not know about the thin and the thick filaments!). It would be a rather ironical twist in the story if the essence of Galvani's ideas turned out to be right after all!

Bibliography

Shelley (née Godwin) Mary (1818). *Frankenstein or the Modern Prometheus.* The 1992 publication of *Frankenstein* by Penguin Classics, is recommended because of the beautifully researched introduction by Maurice Hindle.

Squire, J. (1981). *The Structural Basis of Muscular Contraction.* Plenum, New York.

References

Huxley, A. F. and Niedergerke, R. (1954). Structural changes in muscle during contraction. *Nature*, **173**, 971–2.

Huxley, H. E. and Hanson, E. J. (1954). Changes in the cross-striations of muscle during contraction and stretch and their structural interpretation. *Nature*, **173**, 973–6.

Kabsch, W., Mannherz, H. G., Suck, D., Pai, E. F., and Holmes, K. C. (1990). Atomic structure of the actin:DNase complex. *Nature*, **347**, 37–44.

Rayment, I., Rypniewski, W. R., Schmidt-Base, K., Smith, R., Tomchick, D.R., Benning, M. M., Winkelmann, D.A., Wesenberg, G., and Holden, H.M. (1993). Three-dimensional structure of myosin subfragment-1: a molecular motor. *Science*, **261**, 50–8.

J. BORDAS

Born 1944 in Spain, gained his physics degree at the University of Barcelona and continued his education in France and UK. 1972 completed his Ph.D. at the University of Cambridge and began research on

Synchrotron Radiation (SR) technologies, whose usefulness for research in many scientific disciplines was just emerging. Today continues to work with SR, and has developed many of the techniques now routinely used for research in biophysics, physics, chemistry and materials science. His personal research interest concerns the understanding of the relationship between molecular structure and motion in biological systems. Has spent significant periods of time in many different countries, but mostly Germany where he headed the European Molecular Biology Laboratory Outstation at DESY (Hamburg) and, since 1983, the UK, where he is currently Head of the Research Programme at the Synchrotron Radiation Source at Daresbury Laboratory. Holds a Visiting Professorship in Biophysics at the University of Liverpool.

Risks, costs, choice and rationality

COLIN BERRY

In one of Isaac Asimov's novels, he deals with an 'impossible' crime, the murder of a human being by a robot. The book is a device to allow the author to develop his arguments about the so-called 'Laws of Robotics' but it deals with a society which has become so risk averse that personal contact between individuals is avoided because of the risk of infection—the race is continued by artificial means. The impossibility of the crime depends on the acceptance by the entire population that it is unthinkable that person to person contact would take place, the 'risk' would be unacceptable. In our times, as life becomes more and more secure, we behave like the people in the novel; there is an increasing reluctance to accept risks, however tiny. I believe we are now so concerned about low level risk that we use our judgement poorly in determining priorities in the field of public health and thus affect adversely the well-being of the community at large.

Whatever we are doing to our environment is not affecting life expectancy adversely (Table 1). It is clear that in many societies life becomes more precious as it becomes more certain. We wish to control all aspects of our day to day existence and to make them risk free; we define environments, at work, at home or in the natural world in a way which preserves our view of how they ought to be on the grounds of

Table 1. Telegrams sent by HM The Queen to subjects reaching the stated age in the first and fortieth years of her reign.

	100 years	105 years
Year 1	200	10
Year 40	2227	262

safety, often without having considered what we mean by safety or accept as safe. We indulge aesthetic expectations without close analysis of what is needed to support them—Cumberland should look like it does for ever and the necessity for sheep may or may not form part of the understanding of how that countryside came to be in its current form. At the same time as feeling a need for control we may also dispense with concerns; if we believe that a particular set of activities are desirable or 'healthy' we will accept remarkable burdens of risk in adopting these beliefs, often without information on the extent of potential benefit. Increased mountain walking in the USA has been accompanied by a recrudescence of Rocky Mountain spotted fever (which can be fatal) and the numbers of deaths in mountains in the UK in winter rises as climbing in that season becomes a leisure activity.

In general, populations are prepared to accept risks which increase death rates per year by about one in a million. This is greater than the chance of being struck by lightning (or of winning the lottery) but much less than the chance of death on the road or from natural causes—indeed for much of our lives we are at what is for many people a surprisingly high annual risk of death (Fig. 1 and Table 2). Perceptions of what events or activities are dangerous is faulty in most populations and differs in groups of different ages, educational background and occupation [1] but this is not surprising, even those directly concerned such as members of the medical profession are often uninformed about what types of

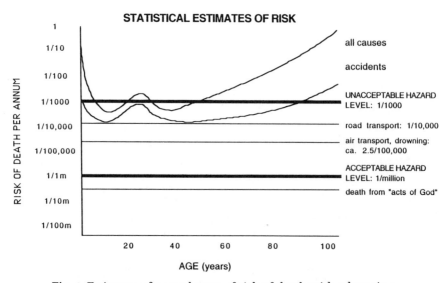

Fig. 1 Estimates of annual rates of risk of death with advancing years and their 'acceptability'.

Table 2. Changes in death rate with time

	AGE	Unicohort for 1 death/year	
		M	F
USA 1990			
	0	75	95
	5	2326	3030
	10	4167	5263
	15	943	2273
	20	541	1639
	25	559	1613
	30	625	1449
	60	52	104
	80	10	17
UK 1992			
	0–1	266	359
	5–9	1883	3141
	10–14	641	1415
	20–24	478	1261
	25–29	472	1150
	30–34	398	790
	35–39	282	505
	40–44	198	302
	60–64	24	38
	80–84	5	4

The unicohort can perhaps be best described as the number of people you would have to follow for a year in order to note the death of an individual. The table shows US and UK figures for different periods. This is deliberate; despite this and the different modes of collection of data, the pattern of mortality by different ages and by sex is closely comparable in two different societies.

accidents cause patients to present in accident and emergency departments if they do not work there. We also accept voluntary risks much more readily than involuntary ones (probably between 1000 and 10 000 times more readily).

In day to day life we continually confuse *hazard* and *risk* and do not pay much attention to the difference between them, but the difference is important and I shall emphasize it now. The seas which surround our islands and the mountains which decorate them are *hazards*—places of potential harm to man. It is not until representatives of our species go and do things in or on them that they become places where *risks* can be identified. The same is true of deserts, rivers, flying and so on—the concept of risk is essentially a quantitative one. So many people per thousand will be killed at sea, in the mines, on building sites in a year, say;

the risk will usually be defined in terms of how many adverse events occur in some stated unit of time (to divers per year or hours under water), distance (so many killed or injured for a given number of road or air miles travelled), or season (numbers lost in Scottish mountains in winter). Hazard identification is easy and often speculative, risk evaluation is often complex and demanding. It is also difficult to decide what we mean by 'safe'. Siddall [2] defined it in the following way: 'Safety is the degree to which temporary ill health or injury, or chronic or permanent ill health or injury, or death are controlled, avoided, prevented, made less frequent or less probable *in a group of people.*' The concept of a *group* is enormously important, responsibility for the prevention of malaria gives you a different perspective on DDT than the view you hold if you look after birds of prey. There is a tendency to ignore the consequences of 'negative' regulatory actions but a ban on the use of herbicides which resulted in the use of more labour and machinery would probably have an adverse outcome in terms of farm workers' health; 61 people were killed on farms in the UK in 1989, mostly by machinery. How do you evaluate the risks/benefits of fruit and vegetables in a diet, the effect of yields on their costs and consumption, the existence of natural toxicant in some untreated foods, the benefit of additives in preventing free radical production—and all these in terms of the individual? I believe we must take a different approach and work from an understanding of mechanisms rather than an extrapolation of estimates.

If we begin from the human rather than the experimental end of things we can consider some real problems of our handling of risk estimation. Lice of several varieties, bed bugs, fleas, mites and ticks all spread disease in man. Leave aside the self-induced problem of trekking holidays in the USA and Rocky Mountain spotted fever and consider the day to day problem of head lice. These insects are commonly found in schools; most parents will have had to deal with them. The conventional (and effective) treatment is to shampoo the hair with a preparation containing HCH (γ-hexachlorocyclohexane) but concerns about the use of this compound as *Lindane* in remedial wood treatment in the home have led to newspaper articles suggesting links between exposure and various blood dyscrasias, notably aplastic anaemia [3]. The Advisory Committee on Pesticides (ACP) of the UK has examined published and unpublished accounts of cases with apparent association of ill health and HCH exposure and has found no evidence of linkage; examination of a number of large aplastic anaemia registries has failed to show an association of the disease with pesticide exposure [4]. It is interesting to note how this question was treated by the media, with emphasis on case reports, often without verification of exposure (something which is very difficult to do—think of Agent Orange) and with causation assumed for widely disparate clinical syndromes of differing pathogenesis including aplastic anaemia, Hodgkin's disease, paralysis and epilepsy. Curiously, and in my view typically, these effects are generally attributed to very low exposures of the kind which might obtain in a house treated by professionals, or to transiently high exposures which might be experienced by a DIY exponent. The effects of the much greater exposures from antilouse treatments or agricultural operations where the compound is used as a spray have been largely ignored. They are seen by the public as being somehow different, just as therapy is seen to differ from passive exposure.

Here is another fundamental difficulty which confronts us in dealing with the generally irrational responses of the partly informed; the willingness to equate *association* with *causation*. None of us thinks rationally without effort or instruction; Hume thought that rationality was a suit of clothes that man put on for special occasions. Figure 2 shows a tempting case (more tempting than many which are assumed to be proven) where the prevalence of sparrows and sparrowhawks appears to show a linkage; the assumptions which might be made are that hawks are the only cause of sparrow mortality and that hawks only eat sparrows. In fact many variables affect this set of outcomes and the analysis of the reasons for change in either species is far more complex.

Nowhere is the problem of irrational linkage more important than in the public perception of the risks of developing cancer. Despite many ill-informed assertions to the contrary [5], there is no evidence that age corrected rates for cancer are rising. In England and Wales, two cancers show an increase which is significant, carcinoma of the lung in women—for reasons sadly known to all of us and ignored by so many—and tumours of the skin which are roughly doubling every decade owing to our collective desire to tan twice a year. Other tumours (cancer of the stomach, Fig. 3) are declining in incidence. But this area of irrationality may now be approached in a more sensible way as the science improves and we may be able to present data which are reassuring to the public. Let us consider this in more detail.

The autonomous cell proliferation which characterizes neoplasia may be due to the removal of a growth inhibitory gene product or the induction of a specifically determined growth rate advantage in a cell lineage. Growth factors which are vital in early embryonic development are produced by genes (proto-oncogenes) which if altered can over-produce normal gene product or altered product with disastrous consequences for the organism (oncogenes). It is important to prevent this damage and highly conserved mechanisms of DNA repair exist to prevent it; interference

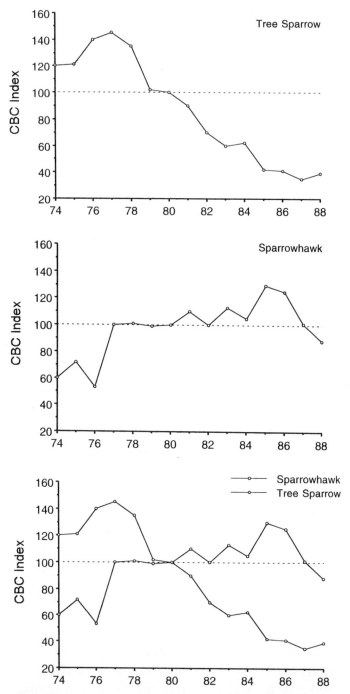

Fig. 2 The figures show what appears to be a clear link between the numbers of sparrows and sparrowhawks or might tempt you to that conclusion, especially when the slopes are combined. The vertical axis represents a ratio of the change of number observed when compared with 1980. After Marchant *et al.* [26].

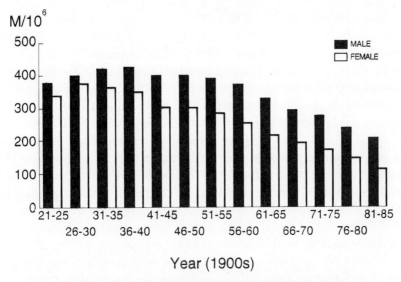

Fig. 3 Gastric cancer mortality, England and Wales. This fall, seen in both sexes, suggests an environmental change, perhaps to do with better food preservation and storage and the disappearance of fungal toxins from food.

with repair may have catastrophic consequences—as seen in the disease Xeroderma Pigmentosum. You and I have genes which act each day in our skin to repair the damage caused to the basal layer of the epidermis by ultraviolet light. In most of us this repair is effective and skin cancer occurs only in those groups occupationally exposed to excessive ultraviolet (farmers, deep sea fishermen) or to fair skinned races living in the tropics (notably in Queensland, Australia). That this mechanism can be exhausted is seen by the fact that high exposures and age are the two major determinants of the rates of skin cancer and that the effect of sunbathing on a normal population is to lower the age at which skin cancer occurs. However, in Xeroderma Pigmentosum the absence of effective repair in the skin means that skin cancer occurs at very young ages and most patients are dead by 35–40 years of age, ages by which very little skin cancer will have occurred in the population at large [6, 7].

Repair of damage to DNA is probably a central issue in many types of cancer in man. There is a link between the sites in DNA which are susceptible to radiation damage and those found to be mutated in cancer cells but this, in the past, has not been found to be consistent. Some easily damaged areas are seldom altered in tumour cells and some infrequently damaged sites are often mutated. However, when repair of the damage is taken into account and particularly when the rate of repair is considered, we see that 'hotspots' for mutations associated with cancer coincide with sites of slow repair [8]. This is important since once a cell divides with damaged DNA the damage becomes fixed in its progeny; further repair of that particular injury will not occur. It can be shown that in many human tumours in genetically normal individuals, genes involved in DNA repair and monitoring are altered. An important part of the process is the gene p53 altered in more than half of all human tumours and in more than 80 per cent of cases in some neoplasms. p53, a cell cycle related DNA binding protein which regulates transcription, acts as a tumour suppressor gene and in many human tumours one allele is mutant and the other deleted [9]. The gene product acts as a cell cycle check point leading to S-phase delay in cells with genetic damage, thus allowing more time for DNA repair before changes are fixed by the next cell division [10]. Levels of p53 appear to be regulated by differential gene expression as well as at the post-transcriptional level and there is also evidence for a particular role for this gene in differentiation; it has been shown that its level of expression in fetal mice is not well corre-lated with cell proliferation but up-regulation of p53 gene expression may be necessary to inhibit cell cycle progression and to allow terminal differentiation at many sites [10].

These advances in our basic understanding have facilitated the evalu-ation of the role of human exposure to chemicals in carcinogenesis in a different way; we now know where to look for damage and may derive information from the type of damage we find. Figures 4 and 5 which I have from Dr N. Lemoine (see also Lemoine *et al.* [11]) show the pattern of mutation found in human colon, pancreatic, and lung cancer. These different patterns indicate that if we assume that particular classes of chemical are likely to produce particular types of damage, then the

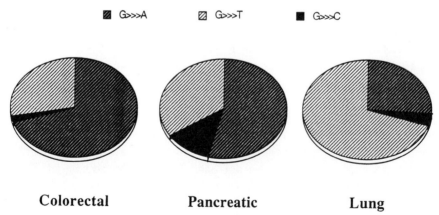

Colorectal   **Pancreatic**   **Lung**

Fig. 4 K-ras mutations in colorectal, pancreatic and bronchial carcinomas.

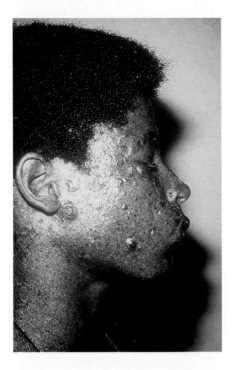

Plate 29 Young male with xeroderma pigmentosum; basal cell carcinoma, squamous carcinoma and melanomas have all occurred on the head and neck.

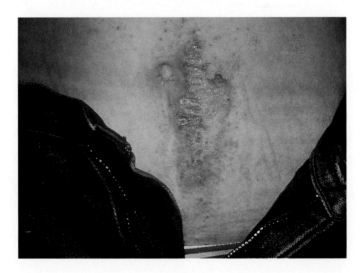

Plate 30 Allergic dermatitis to the nickel in a zip on a pair of jeans. This type of reaction is very difficult to predict and, in these rarer forms, is seen at a very low frequency in the population. Thus, penicillin, a most valuable drug, is nevertheless responsible for a number of deaths each year from allergic reactions whilst being enormously valuable for the population as a whole.

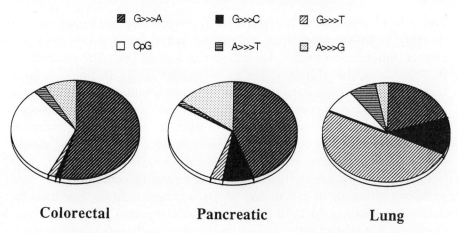

Colorectal   **Pancreatic**   **Lung**

Fig. 5 p53 mutations in colorectal, pancreatic and bronchial carcinomas.

chemicals damaging DNA in two of these tumours are probably similar and the those acting in lung cancer may well be different. The covalent binding of chemicals or their reactive metabolites to DNA produces *adducts* and the extent of formation of these adducts is closely correlated with oncogenicity; as an example, the differences in oncogenicity between isomers of 2,4- and 2,6-dinitrotoluene can be correlated with the extent of adduct formation, the particular adducts formed and their persistence [12]. The methodology of searching for adducts—currently time consuming and technically demanding—is becoming more accessible and the examination of readily available and readily sampled proteins such as haemoglobin offers a field of study which may provide useful population based data.

It is perfectly possible to envisage a situation in which it will be possible to identify the particular causes of cancer within a particular sub-set of the diagnosis. As an example Taylor *et al.* [13] have found that radon-associated lung cancer has a quite specific mutation pattern; in investigating p53 mutations in uranium miners with lung cancer they found that the same AGG to ATG transversion was present at codon 249 in 31 per cent of the large cell and squamous cancers they investigated, including three of five cases in miners who had never smoked. This mutation had only been reported in 1 of 241 published p53 mutations from lung cancers.

But we already are further ahead in allowing this type of information to influence the management of disease. Detailed accounts of the genetic changes in colon cancer, a major cause of death in most developed countries, have led to so much better an understanding of the chronology of events as to give rise to proposals for prevention by once-only sigmoi-doscopy at 55–60 years with subsequent surveillance for the 3–5 per cent of individuals found to have high risk lesions [14]. The rationale is that the sequence of events (six to seven) necessary to produce large bowel cancer will not have time to occur in individuals if early changes are not present at 60 years. The savings over the currently recommended faecal blood/endoscopy regime every 3–5 years are of major significance for the health service, but the knowledge base on which they depend should certainly influence those who regulate risks attributable to chemicals.

Radiation risks present a particular dilemma in combining the two problems of threats of cancer and of the willingness to confuse associa-tion with causation. In 1978 the Boston Globe reported a leukaemia epi-demic at the Portsmouth nuclear submarine facility (Kittery, Maine). Najarian and Colton [15] had examined 1450 death certificates of persons believed to have worked at the yard and who were 'badged' and found six cases of leukaemia where they had expected one. This apparently alarming result prompted a survey of all white males who had ever worked at the shipyard (24 545 men) of whom 7615 had received two rads of lifetime exposure at the naval facility. The complete study showed that the group of exposed workers had slightly fewer deaths from leukaemia than those who had worked at the yard and were unex-posed [16]. This pattern of investigation is also typical of many environ-mental alarms; an initial alarming report is nullified by a careful study—the results of the latter work, however, are often ignored. Similar stories can be told for herbicides and soft tissue sarcoma risks, where a gradual reduction of apparent 'risk' is seen as studies are refined and confounding factors—notably exposures—are considered in detail (Fig. 6). In this instance questionnaires and telephone interviews were used in early (large) studies and were considered to show a strong association, but as various other methodological flaws were identified by critical reviews of the data, they resulted in more and more publications with a steadily diminishing risk. As Gough [17] has said, 'The Swedish studies have been widely cited and probably will be cited for years to come. Peo-ple who cite these studies in arguing for a link between dioxin and soft tissue sarcomas will have to ignore the evidence that doesn't support the link, but that can be done.'

Exposure data are often a problem. Identified individual exposures as a result of monitoring, or after accidents (chloracne after dioxin expo-sures) are hard to find, but even where the precision of estimated values may be comparatively good as, for example, in radiation exposure, the response to the calculated risks is often disproportionate. Let us consider the example of the poorly managed Fernald plant in more detail.

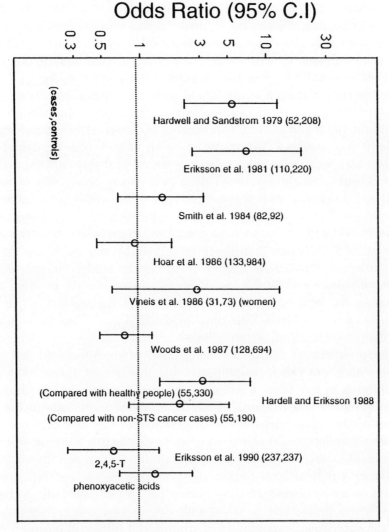

Fig. 6 Case–control studies on soft tissue sarcoma associated with TCDD exposure (modified from Gough [17]).

Uranium is a rather common element, commoner than arsenic, antimony, iodine, mercury, silver, cadmium and about one sixth as abundant as lead—all well known toxins for man. We each have about 60 micrograms of uranium in our skeleton. The radioactivity from this source is not of great significance and the element is regulated mainly by virtue of its toxicity as a heavy metal. The plant at Fernald, Ohio, dealt with uranium as part of US national defence operations and was dealing with large amounts of the element. In 1984 a filter failure occurred and an estimated 124 kg of uranium as dust was released into the atmosphere. Following this release an investigation into the previous history of the plant was made and a truly remarkable story emerged [18] with estimated releases of a total of 179 000 kg of uranium in the years 1951–87 and 6500 kg thorium between 1962 and 1978, together with extensive contamination of the Miami river. This appalling history quite properly led to a study of health effects in the exposed population both within the plant and around it. A number of conservative estimates were made (it was assumed that people spent all of their time outdoors) and particle size was considered in a way which probably inflated estimations of dose, but the estimated dose to a person living closest to the plant and downwind of it was such that they received approximately 40 per cent of what the average US citizen receives from natural sources in a year from the plant. In the same period radon exposures from the plant were around 1000 less than those from natural sources. Using the US Environmental Protection Agency (EPA) linear model for risk assessment, one discovers that the loss of life expectancy for the exposed individual is six days—compared with 2000 days for smoking a pack of cigarettes per day. I have little time for the EPA model but the comparative figures are illustrative, nevertheless.

The problems of the poor management of this plant do not need emphasizing, nor can it be supposed that this degree of environmental pollution is in any sense tolerable, but what followed the identification of the problem was irrational. Bear in mind the magnitude of the risk estimate; an independent study suggested that the loss of life in the exposed population was an excess of seven deaths (0.2 per year) from the 34 years of exposure. Cohen points out that for comparison the air pollution from a typical large coal burning electric power generating plant results in an estimated 25–100 excess deaths per year [19]. Although these figures should also be taken with a pinch of uranium they indicate that the subsequent expenditure of around $1.1 billion in clean-up costs and almost $80 million in health care activities was foolish; it would in no way affect outcomes. However, that part of the money spent on improving the plant ($356 million) was clearly a necessary expense.

As a further example of the disproportionate response, in 1989 a '60 Minutes' programme on American television described daminozide (Alar) as the 'most potent cancer causing agent in our food supply'. There followed a most extraordinary collective hysteria about the supposed risk this compound presented to children in particular, supported in the main by ill-informed and often unqualified opinions from many who were clearly unaware of the fact that there was more hydrazine (the putative carcinogen) in a helping of mushrooms than anyone was likely to get from apples or apple juice. There was no effective response from the United States Environment Protection Agency, actresses became the preferred communicators on the issue. My own calculations of 28 000 apples per day as the danger level, if one assumed that the mouse splenic haemangiosarcomas produced in high dose animal studies were relevant to man, was matched by a calculation by the *Washington Post* that 19 000 litres of apple juice would be the critical intake [20]. It did not help, I was asked to discuss the matter with Pamela Stevenson on Channel 4, a clear indication of national concern. The scientific issues were never discussed seriously in any forum before supermarket chains responded by withdrawing Alar treated apples from sale—a later WHO/FAO view that higher levels of Alar residues than those previously adopted would be safe did not save those apple growers who had been left with an unsaleable crop—one grower in the USA lost part of a farm his family had worked since 1912 owing to consequent financial pressure [21].

The generalization to be made is that rationality is our only defence against what must be regarded as outcomes which are damaging to society as a whole. The difficulty is that we do not behave rationally except on special occasions or unless we are in a framework of thought or a particular working environment where we have learned, often through a time consuming training, to do so. But there are well considered ground rules which we should apply as a matter of course and many media alarms could be avoided if, for example, the thoughtful criteria which Austin Bradford Hill used before inferring a causal association between disease and exposure to putative environmental factors were examined critically before publication [22]. This is not exclusively a media or interest group problem, however, there have been a number of premature associations 'identified' in the medical literature in recent years—these have also ignored the Hill criteria (Table 3).

The consequences of irrational concern will include the irrational use of resources. Table 4 shows the calculated cost effectiveness of selected US regulations according to Belzer [23]. In the upper part of the table it is clear that costs were included in the decision making process, though surprisingly they were not included for those items in the lower part. There are important parallels in Europe in the failure to consider costs in the European Union's stringent drinking water standards. In fact it is rather worse than that; the permitted levels for pesticides in drinking water were decided on arbitrary levels determined by certain analytical possibilities, and toxicological data were not considered. It is estimated that the Water Industry in England and Wales will have invested £1 billion on capital works for advanced water treatment in the period

Table 3 The Hill criteria (Hill [221])

---

1. **Strength of association.** There was a very large increase in scrotal cancer in sweeps.
2. **Consistency.** The effect should be observable in different places by different people at different times.
3. **Specificity.** If the association of a specific type or site of disease is limited to particular groups of workers, and there is no association with other modes of dying—causation is implied.
4. **Temporality.** It must be clear that there is a biologically credible temporal relationship between exposure and effect.
5. **Biological gradient.** By this Hill meant a dose–response relationship.
6. **Plausibility.** The effect should be biologically plausible but this cannot be demanded; what is plausible will depend on the state of knowledge.
7. **Coherence.** The cause and effect proposed should not conflict with what is known about the natural history of the disease.
8. **Experimental evidence.** Can be powerfully supportive.
9. **Analogy.** It is fair to judge by this in some circumstances; the information that Rubella and Thalidomide produce malformations increases the suspicion that other drugs and viruses may act in the same way.

1989–97, mainly in ozone and activated carbon plant. Severn Trent Water alone will have spent around £100 million on capital works and the estimate of their annual running costs is around 10 per cent of the capital costs [24]. The monitoring costs imposed by the drinking water regulations are around £5 000 000 annually, on a national basis. Whatever the numbers, diversion of massive resources to deal with non-existent problems will mean that they cannot be deployed elsewhere, although it is probably naïve to suppose that if money is not spent on achieving unreasonable emission standards it will necessarily be spent on health care, research or other favoured alternatives.

However, there are perhaps more important consequences which follow from the inadequate handling of inconclusive or inaccurate data. By far the most important of these is the danger that where the association of a particular exposure, pattern of life style, industrial process or contaminant with any putative noxious agent or factor has been *mistakenly* identified as causative, action to remedy the situation will be ineffective. The problem will remain, causing damage or injury, when the inappropriate response has been made. An example will make this problem

Table 4. Cost effectiveness of selected US regulations

Regulation	Cost per premature death averted (SM,1990)
Car seat belt standards	0.1
Car fuel system standards	0.4
Car side impact standards	0.8
Car rear seat belt standards	3.2
Asbestos ban	110.7
Ethylene dibromide drinking water standards	5.7
1,2DCP drinking water standard	653.0
Atrazine/Alachlor drinking water standard	92070.0
Wood preserving chemicals, hazardous waste lighting	5700000.0

Data from Belzer [23]. The benefits measured are of deaths or injuries averted, no attention is paid to concepts such as water quality or environmental benefit.

clear; in 1987 the number of cases of listeriosis reported to the Public Health Laboratory Service (PHLS) and its Communicable Disease Surveillance Centre (CDSC) increased markedly. This was a true increase, although diagnostic techniques had also clearly improved over time. The findings led to advice about probable food-borne risks (from cook–chill procedures and unpasteurized cheese) but although there was a clear benefit in terms of reduction in the number of cases, figures remained high until pâté was also identified as a risk factor (Figs 7 and 8). The subsequent reduction in prevalence suggests that this source may have been a major component of the risk in any case. This was a well managed investigation and the 'delay' in identifying a missing factor was short, but the costs of altering food handling and provision was high and

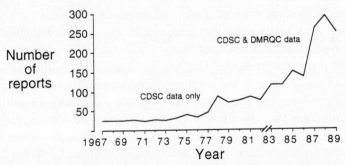

Fig. 7 Listeriosis 1967–89, laboratory reports for England, Wales and Northern Ireland.

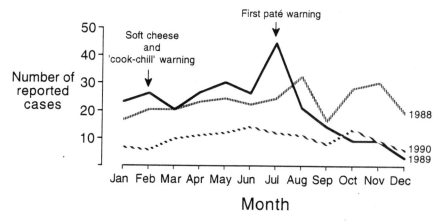

Fig. 8 Listeriosis 1988–90, reports to CDSC and DMRQC for England, Wales and Northern Ireland.

cases would have continued at an unacceptable level despite these expensive manoeuvres.

This example, and that of Daminozide, identify another major difficulty. The measures taken to combat imaginary problems not only fail to deal with real issues but may be very damaging to those whose lives, businesses, prospects of employment or homes depend on particular activities. Moving populations away from non-existent risks is an enormously destructive event, in human terms, but these costs are seldom counted against the activities of those who promote single issues with little attention to the overall picture of community activity, employment or disruption of schooling. The saga of 'Love Canal' where alleged—but not demonstrated—increased rates of cancer, respiratory ailments, nervous disorders, liver damage, chromosome damage and birth defects, around a 'toxic' chemical dump has resulted in actions which caused marriages to break down as families were relocated, should be in all our minds as an example of hysterical overreaction to a trivial threat [25]. This is not a plea for the curtailment of pressure group activity; pressure groups are valuable in identifying issues for evaluation, but they are seldom the best evaluators of community risk.

Is there a general message? Can we draw any conclusions which lead us forward rather than to the depressing viewpoint that it is all too difficult—we must accept the irrational responses of a largely uninformed public and allow that resources will be misused. I do not find this tolerable; it must be the responsibility of all of us to inform where we can, explain the consequences of processes where explanations are possible, seek mechanisms of action of toxic compounds so as to allow real evaluations of risk to be made and ensure that natural phenomena such as radiation lose their aura of dread. This will be difficult; it has taken a great deal of effort to ensure that the general view of smoking, of driving while drunk, or of wearing helmets on motorbikes is a rational one, and some compulsion has been necessary. We cannot compel people to ignore what they regard as risks, but we can and must provide a rational basis for debate.

References

1. Upton, A. C. (1982). *Scientific American*, **246**, 41.
2. Sidall, E. (1980). *Risk, Fear and Public Safety*. Atomic Energy of Canada Ltd.
3. Honigsbaum, M. (1991). *The Independent*, 12 December.
4. Ministry of Agriculture, Fisheries and Food (1992). *Gamma-HCH, a review of the evidence concerning a possible link between exposure to Gamma-HCH and the subsequent development of blood dyscrasias, particularly aplastic anaemia*. MAFF, Pesticides Safety Division, London.
5. Fonda, J. (1981). *Jane's Workout Book*, p. 238. Simon and Schuster, New York.
6. Kraemer, K. H. (1980). *Archives of Dermatology*, **116**, 541.
7. Kraemer, K. H. (1987). *Archives of Dermatology*, **123**, 241.
8. Service, R. (1994). *Science*, **263**, 1374.
9. Weinberg, R. (1991). *Science*, **254**, 1138.
10. Lane, D. P. (1992). *Nature*, **358**, 15.
11. Lemoine, N. R., Hall, P. A., Motojima, K., Urano, T., Nagata, Y., Shitu, H., Tsunoda, T., Kanamatsu, T., Scarpa, A., Zamboni G., and Hirohashi, S. (1994). In *Atlas of Exocrine Pancreatic Tumors*, (ed. P. M. Pour, Y. Konishi, G. Kloppel, and D. S. Longnecker), p. 257. Springer, Tokyo.
12. La, D. K., and Froines, J. R. (1992). *Archives of Toxicology*, **66**, 633.
13. Taylor, J. A., Watson, M. A., Devereux, T. R., Michels, R. Y., Saccomanno, G., and Anderson, M. (1994). *Lancet*, **343**, 86.
14. Atkin, W., Cuzick, J., Northover, J. M. A., and Whynes, D. K. (1993). *Lancet*, **341**, 736.
15. Najarian, T. and Colton, T. (1978). *Lancet*, **i**, 1018.
16. Rinsky, R. A, Zumwalde, R. D., Waxweiler, R. J., Murray, W. E., Bierbaum, P. J., Landrigan, P. J., Terpilak, M., and Cox, C. (1981). *Lancet*, **i**, 231.
17. Gough, M. (1993). Dioxin: perceptions, estimate and measures. In *Phantom Risk: Scientific Inference and the Law* (ed. K. R. Foster, D. E. Bernstein, and P. W. Huber), p. 249. MIT Press, Cambridge, MA.
18. Cohen, B. L. (1993). The saga of Fernald. In *Phantom Risk: Scientific Inference and the Law*, (ed. K. R. Foster, D. E. Bernstein, and P. W. Huber), p. 319. MIT Press, Cambridge, MA.
19. Wilson, R., Colome, S. D., Spengler, J. D., and Wilson, D. G. (1980). *Health Effects of Fossil Fuel Burning*. Ballinger, Cambridge, MA.
20. Marshall, E. (1991). *Science*, **254**, 20.
21. Whelan, E. N (1993). *Toxic Terror*, p. 194. Prometheus Books, Buffalo, NY.
22. Hill, A. B. (1965). *Proceedings of the Royal Society of Medicine*, **58**, 295.
23. Belzer, R. B. (1992). *The use of risk assessment and benefit cost analysis in*

US risk management decision making. International Conference on Risk Assessment, QE2 Conference Centre, London.

24. Leahy, J. (1993). Personal communication.
25. Whelan, E. N. (1993). *Toxic Terror*, p. 121. Prometheus Books, Buffalo, NY.
26. Marchant, J. H., Hudson, R., Carter, S. P., and Whittington, P. (1990). *Nature Conservancy Council and British Trust for Ornithology*, pp. 67, 220. Tring.

COLIN BERRY

Born 1937, qualified in medicine at Charing Cross Hospital. Became Lecturer and, later, British Heart Foundation Research Fellow at the Institute of Child Health. Appointed Reader in Pathology at Guy's Hospital in 1970, and Professor and Head of the Department of Morbid Anatomy at the Royal London Hospital in 1976. April 1994 became Dean of the London Hospital Medical College, and in July 1994 appointed Warden Elect of the new School of St Bartholomew's and the London Hospital Medical and Dental Colleges and Eastman Dental School. Has published some 200 papers on development, vascular biomechanics and toxicology. Served on various committees, including the Committee on Dental and Surgical Materials; the Committee on Safety of Medicines, and the Advisory Committee on Pesticides. Has been a Council Member, Assistant Registrar and Treasurer of the Royal College of Pathologists, and until recently was President of the European Society of Pathology. He is Chairman of the Board of Anatomic Pathology which determines the pattern of harmonization of medical training within the EC. Was made Knight Bachelor in 1993.

The unnatural nature of science

LEWIS WOLPERT

There is a certain ambivalence towards science which is shown by this quotation from D. H. Lawrence: 'The Universe is dead for us, and how is it to come alive again? Knowledge has killed the sun, making it a ball of gas with spots. Knowledge has killed the moon; it is a dead little earth fretted with extinct craters as with smallpox. The machine has killed the earth for us. The world of reason and science, this is the dry and sterile world the abstracted mind inhabits.' Not exactly a very positive attitude towards science! On the other hand, I think many people in the public look to science, particularly medical science, to cure all the diseases— old age, cancer, Alzheimer's, genetic diseases and so forth. But even here there is really quite a lot of ambivalence. Mary Kenny, when colleagues of mine at the Imperial Cancer Research Fund discovered the gene for sex determination in mice, wrote an article in the *Evening Standard* comparing these scientists with the Nazi doctor, Mengele, who did experiments on humans in the Nazi concentration camps. While this is absurd, there is undoubtedly a certain anxiety about the nature of science and what scientists are up to; genetic engineering provides a good example.

There is also a curious image of science, and I think you have to look at this image in cartoons from Sidney Harris. You will notice that the scientist is always male, he is sort of slightly hunched and he is certainly not very attractive. The media have not done much to correct this image of male, unattractive, boring, uninteresting, obsessional scientists. I personally think it is misleading, and I think the problem of image is a very serious one in the public appreciation of science. Can you imagine, for example, the *Sunday Times* allowing an article to appear which says 'Seances are widely used in university history courses to contact the dead'? Well, the equivalent of this is the following: 'Numerous universities worldwide now accept astrological studies as part of a degree course, particularly in association with psychology.' This is from the *Sunday Times*; it must have gone through a sub-editor who actually believes this to be true.

It is because of examples of this kind that I have become involved in trying to understand such attitudes. C. P. Snow spoke about the two cultures, and I do not want to get involved in whether there are one, two or three cultures, but I think it is a particularly British phenomenon that there is something of a divide between those in the Humanities and those in the Sciences. A very influential quotation for me comes from the distinguished literary critic, Lionel Trilling, who said some time ago when speaking about his lack of understanding of science, 'This exclusion of most of us from the mode of thought which is habitually said to be the characteristic achievement of the modern age is bound to be experienced as a wound to our intellectual self-esteem.' If he were typical it would be quite easy to say, 'Look, all that anti-science that there is out there in the community is simply wounded self-esteem, it's envy. It's shame at not really understanding what is probably the greatest intellectual achievement of our age and of our society.' I think that is certainly an aspect, but it is by no means the whole story. However, it did set me to thinking why so many people who were not scientists were a bit hostile to science, or had difficulty with science. Why did they have so much difficulty really understanding it? And the answer I came up with is that science is a rather unnatural mode of thought. What I mean by unnatural is that it goes against one's natural expectations. The strong statement I want to make is that, if an idea fits with common sense, then scientifically it is almost certain to be false. It is nobody's fault that it is like this; it just so happens that the world, the Universe is not constructed on a common sense basis.

Let me give you a few examples. It seems to me that if you get up in the morning when there is some Sun and you can see the Sun during the day, it is absolutely obvious that the Sun goes round the Earth, and I think you would be a fool not to assume that that was true—that would be the common sense explanation. I used to say that everybody believes that, in fact, it is the Earth that goes round the Sun. This, however, is not the case because surveys done by John Durant show that only seventy per cent of the British population actually believe that it is that way round. I do not blame them; I think that it is quite against common sense to believe the contrary. You may feel very confident as a sophisticated audience that the Earth goes round the Sun. But if I were to ask you what is the evidence, my guess is that you would be quite hard put to come up with a good explanation.

Quite a lot of our so-called common sense about science these days is really by authority. Take the tides: we all say with great confidence that it is the Moon that causes the tides. The Moon attracts the water, and so it is high tide on the side where the Moon is. But why is it high tide over on the opposite side? There is a very complicated explanation for this to do with centripetal forces.

From the moment one is born, one is exposed to moving objects. I shall give some examples now about motion which once again confound common sense. Newton's and Galileo's ideas about moving bodies have been around for over three hundred years, yet many people still have quite a lot of difficulty making predictions as to how moving bodies actually behave. An aeroplane is going to drop a ball (it is not politically correct to call it a bomb!), and the question is where is it going to fall in relation to the aeroplane? Many students who have had courses in physics get it wrong; it hits the ground directly below the plane. Another example is: if I take a book and throw it into the air, what forces are there on this book after it has left my hand? Many people, even those trained in physics, think that somehow after it has left my hand there is an impressed force on it which gets less and then it begins to return under gravity. The correct answer is, of course, that there is only one force acting on it all the time—gravity. If you kick a football, what happens to its acceleration from the moment it leaves your foot? The answer is that the acceleration decreases from the moment you kick it, and I think that is quite counter-intuitive. The reason why it is counter-intuitive, and let me give you what I think is a technical explanation, is that acceleration is about second derivatives, whereas speed is about first derivatives, and it is very difficult to be intuitive about second derivatives. A great deal of science depends on mathematics and, unfortunately, mathematics is not an intuitive subject.

Let me give a few more examples. Imagine that I am in a flat field and I have two bullets, one in a gun and one in my hand. If I shoot one bullet horizontally at exactly the same time as I drop the other, which bullet hits the ground first? Everybody's common sense view is that the one which is shot horizontally will get there after the other one. In fact they both get there at the same time. The rate of fall is independent of horizontal motion.

Let us think about ice. I claim that any sensible person who is asked why their drink becomes cold from the ice would think that it is cold going from the ice to the liquid, whereas it is due to heat leaving the water to go to the ice to melt it. It is no wonder that children at school have difficulty with science. Maybe if you tell people that the world is built on a non-common sense basis, then when they come to understanding science they might be slightly more relaxed about it and will not have the expectation that it should fit with common sense.

With really difficult things like quantum mechanics everyday analogies break down. For the big bang, black holes and so forth, there are no everyday metaphors; those subjects are highly mathematical. Even biological ideas like the theory of evolution do not fit with common sense. Do you really believe deep down that humankind has been achieved by random changes and natural selection. In fact I do believe it; I am still surprised by it and I certainly do not think it is intuitively obvious.

There was a time when I used to say, 'Well, there are some scientific laws or some scientific ideas that do fit with common sense.' One is Ohm's Law, which says that the bigger the resistance you have in a circuit, then the bigger the voltage you require in order to drive a current through it. But someone who taught it said, 'Forget about it. It's not common sense and students have a lot of difficulty with it.' And then I remembered my own difficulty about circuitry. You have a circuit—a battery, a wire and a resistance—and now you add another resistance in parallel with the other one. What will actually happen? Actually the resistance falls, and I think that, unless you have quite a good understanding of what is going on, it is very counter-intuitive. So even really quite simple ideas, in physics for example, like Ohm's Law do not fit terribly well with common sense. Thus, the first point that I want to make here is that the world is not built on a common sense basis.

I once gave this lecture in Cambridge. At the end of the lecture a gentleman sitting in the front row asked if I minded if he made a personal remark. He said he would like to tell me what he wanted on his tombstone, which was, 'He tried to understand economics all his life but common sense kept on getting in the way.' This turned out to be James Meade, Britain's Nobel Laureate in economics.

The whole process of doing science also requires a particular mindset, a particular way of thinking about the world. Probability, for example, which scientists use a great deal, does not fit with common sense. Many people who are not familiar with it believe that, if you have an evenly weighted coin, six heads in a row is less probable than heads, tails, heads, tails, heads, tails. In fact, they are equally probable. How many people in a room do you require in order to have the probability of one half, that is evens, of two of them having the same birthday? The answer is twenty-three. You can make a lot of money, knowing this, by going to groups of fifty and betting evens that two people have the same birthday, and you will do extremely well. Once again it is counter-intuitive.

Problems of scale in science are also difficult. There is a story about a Minister of Science who shall be nameless, who went round the Laboratory of Molecular Biology in Cambridge, I understand, and after a whole day, as he was leaving, he said to the Head of the Laboratory, 'Now, tell me, how big is a molecule? Could I hold one in my hand for example?' It turns out that molecules are very small. There are more molecules in a glass of water than there are glasses of water in all the seas. Even quite simple puzzles which involve quantitation do not fit easily with common sense. You tie a piece of string around the equator 20 000 miles long and pull it tight, and now you increase its length by thirty-six inches, so it is 20 000 miles plus thirty-six inches. What is the gap between the string and the Earth all the way round? The answer is about six inches, and most geography students get this right without actually using $2\pi R$. In fact, I used to find it so counter-intuitive that I used to have to check it before each lecture.

The psychologists, to whom I do not usually give a good press, have done a very fine job, particularly Kaaneman and Iversley, in relation to how people make judgements when they have incomplete information. One of their most interesting ideas is what is called representativeness, that is we tend to give undue weight to the most recently acquired information, or that information which we know best. So, if I ask you if there are more words in English which begin with 'r', or in which 'r' is the third letter, you would be an unusual audience not to plump for the first but, in fact, you would be wrong. There are more words with 'r' as the third letter than with 'r' as the first. Another example which illustrates the same point is this: if you go to a group of people and ask them to estimate quickly what is $8 \times 7 \times 5$, and so forth all the way down to 1. Then you go to another group and ask them to estimate $1 \times 2 \times 3$, up to 8. You will get a much higher estimate from the first group than from the second group. Both groups will estimate far too low; the answer is, in fact, 40 320.

One of the ironies about science is that you can live your life perfectly well without knowing any science whatsoever. You may not be able to fix things or take part in any of the major debates related to genetic engineering, nuclear power and so forth, but you can survive perfectly well. Doing science does involve a rigour and a quantitativeness that is not related to common sense. One point about the rigour is the following: in science you have to be internally consistent. Now nobody who has lived with someone else cannot have had the experience that one's partner, either male or female, is perfectly capable of holding two completely contradictory ideas in their mind at the same time. There are many studies by psychologists in which, if you ask people to set out in detail their beliefs about many things, you will find that several of their beliefs are contradictory. But you can survive perfectly well like that. Unfortunately, when you are doing science you cannot do that. You have, I am afraid, to give up some of your common sense views, and it is quite difficult. I have never met anyone at a medical school, where I have been most of my life, who was not absolutely convinced they knew how to choose the best medical students. All the evidence says that we are very bad at such judgements, and you may know many clinicians who will say, 'Well, the trials say that this drug does not work but in my experience. . . .' Clinical trials and double-blind clinical trials are only about fifty years old; they are a very recent invention. The idea that you can convince yourself of the efficacy of a particular activity is all too easy, in relation to all walks of life.

Science is not the same as technology. It is very easy to conflate science and technology, and for all sorts of reasons, particularly when we come to the applications of science, I want to keep a very clear distinction. Many people do not like this distinction, but the more I talk about it the more I persuade myself that I am correct. Much of what we would think of as technology, that is metal-making and agriculture, goes back about 10 000 years. I claim that that sort of technology owes absolutely nothing to science. The French anthropologist Claude Lévi-Strauss writes that, 'Each of these techniques assumes centuries of active and methodical observation, of bold hypothesis, tested by means of endlessly repeated experiments.' Reading that, you could think that technology involves understanding. I would argue however that, until the nineteenth century, science, and by science I mean understanding in some mechanistic and causal way, rigorous and usually quantitative, played zero role in advancing technology, with the possible exception of navigation. Certainly with regard to metal-making and agriculture, you need understand nothing. I would claim that it is all trial and error learning of exactly the same class that chimpanzees use when they join rods together to bring down bananas from a tree. I do not disparage it; on the contrary, I think that one of the characteristics of human beings is technology, whereas science is not a necessary human characteristic. You can be a moderately good cook yet understand absolutely nothing about what is going on. A recipe for pot-making dating to some several thousand years BC reads like a cooking recipe: 'dip the pot in this glaze, then lift it out, fire it and leave it to cool, inspect the result of the glaze, put it back in the kiln, and so on.' The great Chinese inventions, gun powder, the compass and printing, required no understanding of science whatsoever. The great medieval buildings and churches were built with no understanding of mechanics. They were built on what Professor Hayman calls 'the five-minute theorem', that is you put up one of these buildings, you took away the props, you waited for five minutes and, if it lasted for five minutes, you could assume it would last for ever. It is a very good principle of statics. It has been argued that the Greeks could have built the steam engine. You do not have to have any science to make a steam engine; the thermodynamicist came along later to explain why it worked.

My strongest argument comes from evolution, that you can have amazing technology with zero understanding. No one would deny that elephants are a remarkable technological achievement—consider their heart, muscles and so forth. The elephant is an amazing machine, yet it has arrived by natural selection, by random change, and evolution understands no science whatsoever.

I also have Galileo on my side: 'We are certain that the first inventor of the telescope was a simple spectacle-maker who, handling by chance different forms of glasses, looked also by chance through two of them, one convex, one concave, held at different distances from the eye, saw and noted the unexpected result, and thus found the instrument.' Once the nineteenth century came, things changed dramatically, but even in today's society I think there is an enormous difference between knowing how to do something and actually doing it.

It is highly significant that science only had one origin and that was in Greece; and no other society ever developed a scientific approach to the world. The fact that it only happened once supports my argument that science is an unnatural mode of thought. The first recognized scientist in recorded history was a Greek called Thales who lived around 600 BC. What Thales said was that the world was made of water in different forms. According to the historian G. Lloyd, Thales is the first person in recorded history to stand back from the world and take a detached view of it, and try to understand the physical world in a non-mythical way. Moreover, it was a personal view. And then other Greeks came along and said that the world was made of air in different forms, and the debate started. Why science should have started in Greece is not understood. It was Thales too who started geometry; he said that all circles were bisected by their diameters. He moved measurement to mathematics, making general statements about all circles in the Universe, which was a very important step.

The great Greek scientist was Aristotle, but the trouble with Aristotle was that he thought the world was built on a common sense basis. Therefore, most of his ideas about the world are wrong. His strength lay in logic and internal consistency. He had two genius 'descendants', one of whom was Euclid who, although not the founder of geometry, formalized it, setting up a system of postulates from which theorems are deduced. But the person I want to focus on is my hero Archimedes. I consider Archimedes to be as great as Newton and Galileo, because Archimedes was the first applied mathematician, the first person to apply mathematics to understanding the world, and really in a way the world's first physicist. When he lay in his bath, and worked out the loss of weight of a body when submerged, and leapt naked and shouted 'Eureka!', so the story goes, he had made a monumental discovery. His ideas about levers to which he applied geometry were also astonishing. And just look at the way he writes: 'Let it be granted that bodies which are forced upwards in a fluid are forced upwards along the perpendicular to the surface. . . .' It reads like a first year physics textbook. When Galileo, my other hero, says that he could never really have got anywhere without reading Archimedes, I believe him. I do not believe that he was just following in the fashion of the Renaissance to praise the Greeks in order to bolster one's own position. The astonishing thing is that Archimedes' ideas took so long to flourish, and it was only 1800 years later, with Galileo, that they really took off.

I would argue that all science as we know it comes from the Greeks; Islam makes an enormous contribution, and of course there was the flowering in the Renaissance. I know what you are thinking, however; you are saying 'What about the Chinese?' Joseph Needham has written many volumes on science and civilization in China. However, if you read Needham, he makes it quite clear that the Chinese had very little science. They were wonderful engineers, but they had no geometry and they really never understood anything. They had a rather magical approach to science. Somebody once wrote to Albert Einstein, asking him why it was only in Greece and Europe that science flourished. His reply was as follows: 'Dear Sir, the development of western science has been based on two great achievements—the invention of the formal logical system in Euclidian geometry by the Greek philosophers, and the discovery of the possibility of finding out causal relationships by systematic experiments at the Renaissance. In my opinion, one need not be astonished that the Chinese sages did not make these steps; the astonishing thing is that these discoveries were made at all.' If it had not been for the Greeks there need never have been science, because science did not help anybody until the nineteenth century. Francis Bacon, when he forecast the benefits of science, was rather like a politician making promises, for it took another 200 years for any of the applications actually to be of any use.

Why was it in the West that science flourished? I do not want to give the Church too good a press, but I think that you can make a just-so story that Christianity's concept of a Creator, of laws, of internal consistency, really did play an important role. St Thomas Aquinas tried to turn religion into a science; he tried to marry Aristotelianism with religion. In about the third and fourth centuries there was a theological dispute, the so-called Arian heresy, the question about the material relationship between Christ's body and God's body. The very fact that Christians wanted to have an internally consistent argument about this was, I think, at least a step towards science. If you contrast that with Taoism, the dominant Chinese religion, you see a very large difference. For example, a quotation from the Tao says, 'Don't meditate, don't cogitate, follow no school, follow no way, and then you will attain to Tao.' It may be the right way to spiritual liberation but not the way to do science!

Aristotle said that heavy bodies fell faster than light bodies. You all know the story about Galileo who went to the top of the Tower of Pisa and dropped two bodies, one heavy and one light, and they reached the ground at the same time. In fact, the experiment had been done about 1000 years earlier. The point about Galileo is that he showed that you did not have to do the experiment; there was a logical contradiction. 'Let's imagine,' he suggested, 'that we go to the top of the tower, any tower, and we drop these two bodies, one ten times heavier than the other, and that Aristotle is correct. The heavy one gets there faster. Let's do the experiment again, but now we join them together. What will happen? The light one slows down and the heavy one speeds up. But together they weigh more than either, so according to Aristotle they should fall even faster!' It took something like 1800 years for someone to realize this internal contradiction, which supports my argument rather strongly that science is not built on a common sense basis.

Turning to creativity in science, Jakob Bronowski has written: 'The discoveries of science, the works of art are explorations, more are explosions of a hidden likeness. The discoverer or the artist presents them in two aspects of nature and fuses them into one. This is the act of creation in which an original thought is born, and it is the same act in original science and original art.' It is a widely held view among scientists in particular and maybe among some artists that the act of creation in science is very similar to that in arts. I think this is sentimental nonsense and I want to try to tell you why. There are certainly similarities between creativity in any field but I have never heard any scientist say that his creativity in science is just like that of a creative accountant! By conflating the two, one loses the important differences between the arts and the sciences. Science is progressive; science approaches closer and closer to understanding the nature of the world. You cannot talk about art being progressive; art is about change. But there are many other differences: a work of art has a high emotional content, the original production is the crucial thing, and it has multiple interpretations.

Compare that with Newton's discovery of the calculus; nobody, unless they were a historian, would go back and read his description of the calculus—it is very difficult. The ideas have been incorporated into mathematics in a much more accessible form. *Hamlet*, by contrast, does not become amalgamated with *Arcadia* in the same way that science is put into a common body of knowledge.

David Hilbert pointed out that the measure of a good scientific paper is how many other scientific papers it makes irrelevant. When Tom Stoppard wrote *Arcadia* it did not in any way do away with *Hamlet*. Moreover, if there had not been Shakespeare, nobody would have written *Hamlet*. Ultimately, however, all scientists are anonymous and irrelevant; if it had not been Faraday, it might have been someone else. With the DNA story, if it had not been Watson and Crick, we know it would have been Franklin and Klug within a year. We know how much simultaneous discovery there is in science; you do not have simultaneous novels or simultaneous paintings. It is good to see the similarities between art and science, but not to see the differences actually confuses the whole issue. Of course, there is a great deal of creativity in science, but of a rather different kind to that in the arts.

I dislike the concept of serendipity. I feel it is used almost to disparage science. They said of Fleming, for example, 'Isn't he lucky!', when he discovered penicillin, yet he spent most of his life looking for something like an antibiotic. We could say, 'Wasn't it lucky that he was born! Wasn't it lucky that there was St Mary's Hospital Medical School! Wasn't it lucky that he had parents at all!' Life is full of contingent events, and I am always struck that it is the best scientists who are the 'luckiest'. When Pasteur was told that he had been very lucky, his famous, and probably irritable, reply was: 'In the field of observation in science, fortune only favours the prepared mind.'

Having recognized that science is peculiar you might have thought that either the philosophers of science or the sociologists of science, those professionals who devote their time not to doing science but to thinking about science, may have illuminated its nature. The philosophers of science this century, however, have made virtually no contribution to the understanding of science. I am a crass, naïve realist. This is, philosophically, totally indefensible but it does not matter. You can still do science perfectly well by taking such a position, even though you cannot defend it. What about Karl Popper? Unfortunately, I regard Karl Popper as one of the great overrated philosophers of science. I think Popper's great idea that the only way science progresses is by falsification is itself grotesquely flawed, and the reason why it is grotesquely flawed is explained in this quotation from Francis Quick: 'A theory that fits all the facts is bound to be wrong, as some of the facts will be wrong.' How does one know a falsification is correct? Moreover, the whole idea fails to deal with discovery. Science is a very complicated process; there is no algorithm for doing science, no general scientific method. Peter Medawar made it very clear: 'If the purpose of scientific methodology is to prescribe or expound a system of inquiry, or even a code of practice for scientific behaviour, then scientists seem to be able to get on very well without it.'

The sociologists of science are a somewhat different group. There is a group who support what is called the Strong Programme in the Sociology of Science, who take a very curious view of science. Science, they say, is assumed to be the form of knowledge which *par excellence* remains unaffected by changes in social contact, culture and so on. This recent social study of science, they claim, challenges this assumption. Their argument now is that the universalism of scientific truth is a myth. We should abandon the idea of science as privileged or even a separate domain of activity and enquiry. For some of them, the natural world has a small or non-existent role in the construction of scientific knowledge. You may think that this is some Monty Python story; it is not. Such ideas are widely taught at British Universities by sociologists of science in courses on science studies. Relativism is rampant among sociologists of science. They really do believe that all science is a social construct. I wrote recently in the *Times Literary Supplement* that nothing coming from the Strong Programme is not either obvious, trivial or wrong, and I still hold this view.

Leo Tolstoy has written: 'Science is meaningless because it gives no answer to our question, the only question important for us: "What shall we do and how shall we be?" ' And I have to say that I agree with Tolstoy. I think that science has very severe limitations—it does not tell you how to live your life; it has zero to say about ethics and moral issues. Once you know what you want to do you can apply science and that can help you. I would also argue that religion and science are totally incompatible. However, people can be deeply religious and outstanding scientists at the same time. This may be because religion is a natural mode of thought and science an unnatural mode of thought, and the two can live very happily together. Faraday, a genius scientist, was a deeply religious man, and I would say that if you took a survey of scientists in this country you would find that more than fifty per cent of them at least were very religious indeed. My position, following David Hume, is that science is about reason and religion is about faith.

But what about scientists and their social responsibility? There is at the moment a great deal of anxiety about scientists being socially responsible.

The big issues are nuclear energy, the weapons and genetic engineering. I suggest that the only social responsibility scientists have, because they have access to privileged knowledge, is to explain to the public what the implications of that knowledge are and how reliable that knowledge is. It is not up to them to make the decisions. I would argue that, in relation to the bomb, the scientists behaved with great moral responsibility. Robert Oppenheimer made the position very clear when he said: 'The scientist is not responsible for the laws of nature, but it is the scientist's job to find out how these laws operate. It is the scientist's job to find the ways in which these laws can serve the human will. However, it is not the scientist's responsibility to determine whether a hydrogen bomb should be used. That responsibility rests with the American people and their chosen representatives.' On the other hand, I would argue that the scientists committed to eugenics in the early part of the century behaved rather badly. I would argue that the eugenicists, in the early part of the century, not only did not make the social implications of their work clear, and discuss their reliability, they actually perverted their science in order to put forward a particular social programme. Konrad Lorenz even made an analogy between bodies and malignant tumours, on the one hand, and a nation and individuals within it who have become asocial because of their defective constitution.

Where do we stand today then with issues related to genetic engineering? People have always had a fear of chimeras. I sometimes say that Mary Shelley is the unintentional evil fairy godmother of science, because Frankenstein and his monster really pervade a great deal of discussions. In relation to genetic engineering, my views are that the fears are grossly overrated. As a cyclist in London, I claim that potholes in Camden are a much greater social danger than genetic engineering, because the danger is imminent. People actually have very severe accidents with potholes. Genetic engineering and the whole Human Genome Project as yet has harmed no-one, and in principle it is no more dangerous than any other medical treatment. This is not to say that there are not problems. I would say that understanding people's genetic constitution does carry with it the problem of privacy, particularly in relation to insurance, and this is one issue that people are going to have to deal with. People say that there is a tyranny of knowledge that comes with genetic engineering. For example, prenatal diagnosis puts, it is said, enormous stress on people with respect to whether to terminate a pregnancy or not. With knowledge comes responsibility. You can decide not to have the test. Life is filled with difficult decisions. There is, for example, a disease called Huntingdon's Chorea which comes in middle age. It is now possible to have a test which will enable younger people to find out whether they will actually get the illness. What the surveys show is that not everyone takes the test but that people want the test available. My own position is that the decision on most of these things must be left to the individual, not to the doctors and not to the scientists. The social responsibility of the scientists is to make the implications clear and also to point out the reliability.

I am extremely upset by Jill Knight in the House of Commons for attaching a clause to the criminal justice bill which outlaws the use of eggs from aborted fetuses. It may eventually be outlawed but I do not think it is her right to do it. I think this is a delicate area which needs to be brought into the public domain; there should be wide discussion, then if Parliament wishes to take a decision it can do so. What I am against is any expert, anyone with a particularly strong view, actually making the decisions for us. And this brings me to a key quotation from Thomas Jefferson in relation to the public understanding of science, to which I give my strongest support: 'I know no safe depository of the ultimate powers of the society but the people themselves, and if we think them not enlightened enough to exercise that control with a wholesome discretion, the remedy is not to take it from them, but to inform their discretion.'

I leave you, then, with the words of Einstein, which again reflect on the peculiarity of science: 'The greatest mystery of all is the partial intelligibility of the world.'

LEWIS WOLPERT

Born 1929 in South Africa and studied civil engineering at the University of Witwatersrand. Moved into cell biology in 1954, and became Reader in Zoology in 1964 in the Department of Zoology, King's College London. 1966 appointed to his present post. 1984–8 was a member of the Medical Research Council and is currently Chairman of COPUS. 1986 gave the Royal Institution Christmas Lectures. 1988–9 was presenter of the BBC 2 science programme 'Antenna'; has conducted 25 interviews with scientists on Radio 3, published in *A Passion for Science*. Publications include *The Triumph of the Embryo*, 1991 and *The Unnatural Nature of Science*, 1992.

Science and political power

TONY BENN MP

This lecture is about science and political power, or perhaps more appropriately, the scientist as a citizen. It is impossible to live your life, even without scientific qualifications, without coming across the impact of science from early morning till late at night. I am reminded of the day I learned to fly fifty-one years ago and on my first solo I was a little nervous. My instructor said 'Don't worry about the landing, we've never left anyone up there yet.'

Most of my ministerial life, extending over many years, has been on the borders between politics, science and engineering, learning about the brilliance of British scientists, the Nobel Prizes that have been won and the progressive process of industrialization. In 1948, forty-eight per cent of all the ships launched in the world were launched in British ship-yards. Now there are some hoping to get a Trident refurbishing contract. We had the largest machine tool industry in the world twenty years ago and the largest motor car industry in Europe. We had the largest motor bike industry in Europe in 1960, larger even than the American. It has all gone. I remember even as a child reading in my books, as everyone will have done, 'Britain becomes the workshop of the world.' Now you would have to find a new title for the chapter.

The impact of science and engineering on our lives is very, very profound. It is a difficult area to talk about, the relationship between science and politicians, a very delicate relationship, because the attitude of many scientists to politics is full of quite understandable suspicion. Politicians deal with subjective prejudices and unprovable propositions, characterized occasionally, I am told, by ambition, a little skulduggery, occasionally some corruption—Nolan is looking at that now—and not a little double-dealing. Scientists feel that they are being drawn into the party political game and they do not like it very much. The true scientists are proud of the fact that they deal with ascertainable facts, with the immutable laws of nature which can be mathematically checked; that they discover and present objective truth; that science is international, which indeed it is. They believe that all learning benefits humankind, and as a result scientists are the servants of humanity and that they must have intellectual freedom to pursue their researches wherever they take them. Wherever scientists go, public funding is absolutely essential in order for science to survive, and educational provision is a part of it. Recently in the House of Commons we had one of those very rare debates on science policy to indicate the importance Parliament attaches to it. Three hours were allocated for the purpose and I was provided with a brief prepared for Members of Parliament by the Royal Society of Chemistry, which drew attention to the fact that UK research and development has declined over the last five years, that Government funding has declined, that the amount of money spent per full-time academic in universities has declined, that short-term contracts have grown. Over the last ten years, whereas there were once 11 500 people on short-term contracts, this has almost doubled to 21 500. These issues were aired in the House of Commons, and I certainly recall as Minister of Technology a tension expressed itself in the sense that many scientists would come and say 'We would like the funding but don't interfere', and 'We want science to enjoy stability without politics'; but all funding involves political decisions.

When, many years ago, I was in Leningrad, now St Petersburg, one of the most senior Soviet ministers said to me 'Science is an expensive way of satisfying your curiosity at the expense of the state.' And I knew what he was talking about. In order to justify my courage, I refer to a quotation which I find very helpful when moving into areas in which I do not have any great expertise. Professor Seligman wrote an article in the 1950 edition of the *Encyclopedia Britannica* in which he used these words: 'Culture is communicable intelligence. The other animals have no culture worthy of the name because their powers of intelligent intercourse are slight and in particular confined to those alive and present together. Man, however, through his culture, can defy time and space, taking counsel with the dead and gathering in wisdom from the ends of the Earth. There lies humanity's chance of eventually acting together, namely in the power of talking out its difficulties and so getting its crossed purposes straightened out.' I've always found that a very moving passage because certainly at The Royal Institution, which has a history going back two hundred years, the capacity to take counsel from the dead is very, very evident all around us, and the capacity, because of the international nature of science, of drawing experience from other people of different cultures, of different traditions, is also at our disposal.

When I came to The Royal Institution in 1969, I talked about the people versus the machines, and I tried to illustrate the nature of the technical change, all very familiar particularly to primary school children, and some of the older people have learned it as well, and what had actually happened within not just our lifetime but within the last few years. I tried to analyse what had happened in terms of the growth of machines behind which all scientists have made their contribution.

Take travelling machines. You begin with how fast you could go on foot, and then by horse, and then we move forward to the stagecoach, and on until the train. Then in 1945 came the supersonic aircraft moving at 700 miles an hour, and now the spacecraft at 25 000 miles an hour.

Or take communication technology. Early people could address those who could hear them physically. Then somebody invented the megaphone, and in the twentieth century you got radio, and then in 1945 you had the international cables, and now the whole population of the world can hear on television what is being said.

Look at killing machines, which are always very important, particularly to governments. In the early days you could stab somebody or strike them with a sword, and then Genghis Khan came along on horseback and improved the delivery mechanisms. Even as late as 1900 the largest number of people you could kill at any one time was the number of people you could dispose of with a drum of ammunition in a Lewis machine gun. And then in 1945 came the atomic bomb, and you could wipe out the population of the world quite easily.

Move on now to calculating machines. First your fingers, then the abacus, Babbage's analytical engine, and in 1945 with computers, millions of calculations could be done in a micro-second.

All this demonstrates two things: one is the generation gap between people born when I was, in 1925, and my grandchildren, who take it all for granted. There are a lot of children who can teach their grandparents as you will know. Secondly, we learn about technological expectations. If this is what you can do, why do most people live at a low level? I remember when the Russians put a space vehicle on the moon, somebody wrote to me and said 'Why can't they have a decent bus service in Bristol?' An extremely serious argument which I had to take seriously.

I also wanted to show when I last lectured at The Royal Institution what those technical changes had done in terms of the growth of organizations, and I took three types of organization: national budgets, defence budgets and what forms the overall turnover of a firm. The largest organization in the world was then the American government, the second largest was the Soviet government, and the third largest were the American defence forces, which cost more than the whole British national budget. General Motors at the time was a little bit bigger than Japan (I doubt whether that is still true), Ford was larger than the French army, and the Ministry of Technology over which I presided was a bit larger than the Norwegian national budget. Scientific advance creates enormous organizations, and it is about them that I should like to talk. I should possibly update some of that information by telling you very simply that since 1969 the world's population has doubled, world energy demand doubles every fifteen years, and world computing power doubles every twelve months, which gives you an idea of what is happening.

What I said on that occasion was that the technological revolution, by disseminating so much to so many people, had begun to break down authoritarianism in the world and the real conflict was between the people and the machines rather than between government and the governed. Now I am not sure that it was true even then, and I am sure that it is not true now. Because it is very, very hard, when you look at the power those human organizations have, to accept that their authority has in some way been undermined. The real revolutionaries in the world are of course scientists and technologists and all the revolutionary changes that have occurred derive from advances in science and engineering. They are having a huge impact, and will continue to have an impact, on every aspect of our life. If I take one of the oldest sciences or technologies, that of the architect, the built environment, the way in which we live and the way in which our buildings shape us after we have shaped them, is of enormous importance. I remember Buckminster Fuller, the American, whom I had the privilege of knowing, saying in a lecture that he gave to the Conference of Architects in Mexico City, 'It is a very strange thing that when the completely built environment was first devised, i.e. the space capsule, they didn't bother to ask the architects'. That was a profound indication of the way in which architecture and some of its fancy styles had moved away from what it really was about, providing some cover for the human race to live.

All these changes have had a very profound effect on the nature of work, since work is becoming increasingly broken down, disaggregated. A lot of people work at home now, back to the old home-working, if you like, where people did their knitting and sold it, but the isolation at work is very profound. Secondly, the effect of technology on the vulnerability of systems. Tear one page out of a book and it probably will not affect your capacity to understand it. If you pull one wire out of a system, the whole thing goes blank; people may not realize the vulnerability that comes with technical change.

Obviously, this has a huge impact on education. I am strongly of the view that you should raise the school leaving age to ninety-five. The first time I made that speech I said ninety, and I had a very angry letter from a pensioner in Liverpool who said 'Dear Tony, I've just done an Open University degree at ninety-two. Will you raise the school leaving age again?' So I have done. Education is not about stuffing people up like turkeys for Christmas. Education should be an escalator going alongside people throughout the whole of their lives. Schools and further and higher education should be like a library; you go in and learn what you want to learn, when you want to learn it, the way you want to learn it, and you come out and you have learned it. The revolution in education, particularly in adult education, is something some people have not fully appreciated.

Science and technology have had a huge effect on government, a huge effect on international relations, and on the prospects of human survival.

So if I sum up what I am saying about the role of scientists, I think that the real revolutionaries are not, as you may think if you read *The Sun*, or *The Mail* or *The Express*, shop stewards, or left-wing activists, or troublemakers, or demonstrators, or parlour pinks from Hampstead, if there are any left. The real revolutionaries are those who are studying and applying the laws of nature, who have literally turned the world upside-down and sometimes seem to want to walk away from the chaos that follows because they do not want to dirty their hands with politics. And that is the area I want to cover. Of course, it was all predicted very clearly by a very famous nineteenth century philosopher who said 'Technology discloses man's mode of dealing with nature, the process of production by which he sustains his life and thereby lays bare the formation of his social relations and the mental conceptions that flow from them.' This beautifully clear explanation was written by Karl Marx. The question that scientists have to ask is this: 'Given the impact of science on society, how can the scientific community play a much more active role in explaining what is happening and why it is happening, in identifying the objectives we should set ourselves, in helping us to find ways of realizing those objectives, in influencing the political decisions that have to be made and, above all, by contributing to the ongoing debate about the proper moral basis for our decisions?'

If you try to draw a balance sheet of the effect of all the changes I have described, it is not difficult to see the immediate advantages which accrue to many people though not to everybody: higher living standards, motor cars, television, washing machines, worldwide access to knowledge. That is certainly becoming truer with CD-ROMs when people will be able to tap into world libraries and have access to them, better health, a longer life, and easier travel.

These are some of the most immediate and obvious advantages of scientific change. But if you look at the other side of the balance sheet, as any serious person has to do, you can also see the most appalling problems which need to be resolved. The one which was most obvious of course in recent years was the risk of nuclear war. I visited Hiroshima and Nagasaki ten or twelve years ago and saw a city utterly obliterated. Of course, it has been rebuilt now. If anyone has ever been to Hiroshima and Nagasaki and been to the exhibition there, it is quite terrifying. The spread of nuclear weapons and the possibility that they might be used by accident is something that hung like a cloud over us. The threat of pollution which is something that particularly concerns younger people, the anxiety about the environment, the ozone layer, the whole question of our capacity to find ways of maintaining a stable and sustainable system of life, and the creation of huge new power structures to which I referred. Industries now operate on a global scale and it is no longer possible for people working in industry even to be sure who owns them, because they may find their own company or factory has been bought by someone else. Multinational companies have enormous power in the world.

Then you have the growth of global finance. I do not know how many people listen as I do occasionally to the BBC business news. I am fascinated by Mr Dow Jones who seems to work endlessly on his industrial averages; I can only admire him for his talent. We are told the pound sterling has dropped three points of decimals against the basket of European currencies. I have never had a basket of European currencies but I am going to take one next time I go on holiday on the continent. It sounds a very handy thing to have. After the pound sterling has fallen, three hospitals are closed and international confidence is restored in the British economy. This is the nature of power. You have to face it. I sat in the Cabinet in 1976 when we were told by the bankers we had to cut £4 billion off our public expenditure. All governments in the world are now effectively controlled in their capacity to deal with problems by the view the market takes on what they have done. This is a product of a whole range of factors: the industrial development of multinational companies, the role of the banking industry, the use of computers that allow you to discover in a second what is happening to every company and every country in the world.

Take another aspect of technological power. Governments have very big budgets that are used for all sorts of other things. The security services have enormous dossiers on everybody. My last remaining link with the British Establishment is through the telephone, so I always talk very clearly and hope that they will understand what I am really saying. In every country in the world telephones are tapped. That is what GCHQ is about. I can tell you from actual experience that we have an arrangement with the Americans, whereby they lend us warheads for our missiles if we will provide them with all the bugging that we do through GCHQ. This is the way the security services operate.

Then you have, of course, the arms trade, which I personally think is more dangerous than either drugs or AIDS because it is publicly sponsored. You know what happens: British arms manufacturers, with the technological capacity they have, supply arms to people. As soon as they are used the British government calls for a ceasefire and then the arms manufacturer goes to see which weapons work best and sells more of them a bit later. That is the use of technology for a combination of commercial advantage and foreign policy. The command over resources that government has, good or ill, is a very formidable one. Look back on the cold war—it did cost us a great deal of money. I was in Parliament as a very young Member in 1951 when Aneurin Bevan resigned from the Cabinet because he thought that the rearmament programme was too great, and he said then that he doubted whether the Russians had the capacity, or the desire, to invade Western Europe. Yet we were told for forty years that in no time at all the Red Army would occupy West Germany, Italy and France and come to London. That was the basis of an enormous expenditure of money for weapons. Then you see how diffi-cult the Russian army found it to try to occupy Chechnya. President Eisenhower in his farewell address spoke about the military industrial complex, the capacity of that organization to get hold of an enormous number of resources, and it is still true.

I do not find my mind moves very easily over the millions and tril-lions of pounds that are referred to in government budgets, so I got the House of Commons research department to break down the national statistics by population. So when I knock on the door of my constituents and say 'How are you getting on?', they say 'Well, Tony, it's a bit diffi-cult.' So I say 'What is it?' and they say 'Well, the money's hard.' And I say 'Well, aren't you spending a bit too much on weapons in this house?' And they say 'We haven't even got an air pistol.' I tell them a family of four is spending £40 a week on weapons. Then I try to cheer them up and I say 'Have you had your cheque for the North Sea oil revenues yet?' And they say 'What do you mean?', and I say 'A family of four, you should have had a cheque for £9563. Has it arrived yet?' A lot of these economic issues are made so big and complicated that I think it is very important that we should try to recover the common sense that is a part of the scientific process.

Technology has also made the media very powerful. The parallel is with the old medieval churches. Our oldest nationalized industry, as you know, is the Church of England, nationalized by Henry VIII, who had some difficulty about his marital arrangements, so he took over the Church of England. But why did he take over the Church of England? Because he wanted a priest in every pulpit, in every Parish, every Sunday telling the faithful that God wanted them to do what the King wanted them to do. When I was Postmaster General I wondered what great radical instinct had made Charles II nationalize the Post Office. I discovered that he wanted to open everybody's letters and the only way he could do it was through the Royal Mail. MI5 began in 1660. It was the Conservatives who nationalized the BBC because they wanted a pundit on every channel every night telling you there was no alternative. The media shape people's minds. The most important thing that we have to talk about, when we talk about these enormous new organizations, is the philosophy that inspires them. What is it that really makes them move? The real political correctness of the world in which we live is an uncritical and slavish obedience to market forces, to a great extent bypassing and undermining the processes of democracy. These rights were fought for over a long period by people who did not have any power but thought if they had access to the ballot box they might collectively combine and have some influence on policy. When you look at the ways in which these ideas develop, you will see what a very, very strong influence they have on society: the commercialization of human relations.

Consider the word 'customer'. You get on the train and the ticket collector says 'Customers who boarded the train at Derby will please look after their baggage.' It is a sensible piece of advice, but when the ticket collector came round I said to him 'Excuse me, I'm not a customer, I'm a passenger.' He looked in his little book from Railtrack and he said 'I'm very sorry Sir, you are not a passenger any more you are a customer.' So I said 'Well, if I go to have a hip operation in hospital, would I be a patient or a customer?' Well, Railtrack gave him no advice on that point whatever. I said 'If I'm pulled up for careless driving, would the magistrate say "Customer at the bar?" And does the Archbishop of Canterbury talk of "our dearly beloved customers"? Does the Queen talk about "our customers here and abroad"?' But of course the thing about a customer which perhaps you will have noticed, is that if you have no money, you cannot be a customer. So the word customer is a word that dehumanizes and depersonalizes the poor. You go along the Embankment in London and you see people sleeping in cardboard boxes. They, probably more than anyone, need and deserve houses. But they are not customers, because they cannot afford them.

This philosophy is not a party matter, because this virus has infected a lot of people. It is something that really does need to be examined because if large organizations are motivated by that criterion it explains an awful lot of things that happen. Every system of government requires some discipline. In Islamic societies, if you steal they chop off your hand. In Moscow in the old days if you caused trouble you were sent to Siberia. Our discipline is unemployment. I have never been of the view that we have an incompetent government. I think it is the most competent government I have ever met in my life because four million unemployed keep people in order. If you lose your job you might not be able to keep up your mortgage, and if you cannot keep up your mortgage you will be repossessed, and you will be in a cardboard box like the people you see on the Embankment. It is a form of discipline that keeps everybody in order. Homelessness plays a part too for that purpose—it is fear. If people on short-term contracts take out a mortgage, how are they going to manage? This idea that anxiety only applies to perhaps motor workers or miners who are sacked and so on is absolutely untrue. There are a lot of people who have very highly skilled professional jobs, who do not know whether they will have that job six months or a year ahead. It is very important to realize that everyone is kept on tenterhooks. I believe that this plays some significant part in the breakdown of the social fabric, the erosion of responsibility in one form or another, crime and drugs and violence and the feeling of hopelessness and pessimism. They are the result of our failure to cope with the consequences of technical change and to see the world in the same way that the over-whelming majority of people see it, which is really quite modest. They would like useful work at a living wage, they would like a good home to live in, they would like lifelong education, the kind to which I referred, they would like healthcare free at the point of use; people want dignity when they are old, some sense of security in life, and peace. If you look at the world that way, rather than from the point of view of the authori-ties I have described who have come to run large organizations, you get a very different perspective about how science might properly be used. These are the issues which I think everybody, scientist and non-scientist, should be thinking about, because in the end it is the objectives that you are trying to pursue that really matter.

In war time it was very easy, you just tried to defeat the enemy. Every-thing took second place to that. Indeed there was no problem of un-employment during the war. I had a letter from the Government when I was seventeen and a half. It says 'Dear Mr Benn, If you turn up on a certain date, we'll give you free food, free clothes, free accommodation, free training, and all you've got to do is to kill Germans.' It solved the problem of unemployment overnight. The Germans had a similar scheme, a youth training scheme, you turned up and they gave you free food, free clothes, free accommodation, free training, to kill the British.

It has often puzzled me why it is possible to end unemployment for the purposes of destruction, and yet not for what needs to be done now. There are houses to be built, rivers to be cleaned, nurses and doctors to be recruited, school books to be printed, schools to be repaired. So why do we not use the human resources at our disposal for the purpose of meeting needs? Unemployment costs £26 billion a year, and that leaves out the amount of wasted production that would be brought about if every one were employed. We spend more on unemployment than we do on defence, so unemployment must be very important to somebody. How did we end the unemployment? We ended it by rearmament. I am not recommending that. They took unemployed people off the dole, they put up new factories to produce ships and guns and tanks, and then instead of being on the dole, people had a wage, they paid taxation, which paid for the war. My mother never had a Spitfire, and my dad never bought a tank, and granny never had a machine gun. There were no market forces in it at all. I read in the paper that one of the effects of the earthquake in Japan was that unemployment would end because they would be using everybody to rebuild the cities. Why can we not do it now? Because it is not profitable to do it. That is the reason it is not done.

I was asked to go to a comprehensive school in Chesterfield the other day to attend a business studies course and when I got there, with about five or six youngsters in the class, a teacher came up to me and whispered in my ear 'Tony, I used to teach social studies but as you know that's illegal, so I'm now teaching business studies.' And I said 'What's the project?' She said, 'Well, I'll tell you what the project is. The project is that the local authority is going to close this school and our business studies course is to maximize the profit from the sale of the school.' She pointed to a rather nervous girl and said 'She's a property developer', she pointed to somebody else and said 'He's an estate agent', and to another 'He's the chairman of a housing company'. And then she pointed to the most miserable lad in the class and said 'He's the head teacher who is trying to improve his redundancy pay'. So I turned to the children and said 'Is it about maximizing profit?' 'Oh yes', they said, they were so keen the teacher would notice. I said 'Are you sure you're picking the most profitable things? Have you considered prostitution, because there are a lot of girls in the school, or what about making flick knives in the school laboratory for football fans, or could you not open the playground to toxic waste from all over the world? You could earn far more than building houses or whatever it was.' The teacher's face fell a mile. I said to the youngsters, 'Do you want to close the school?' 'Oh no, my granny was here, my uncle was here.' I said 'Why don't you drop the whole project and march on London, and keep the school open.' But it was interesting because the indoctrination of those children into the principles that profit was more important than education was as deep and great as any you get in any fundamentalist muslim country.

I am trying to explain what happens. All these problems are challenging, difficult, interesting problems. Yet the level of political dialogue in Britain at the moment is more shallow, more personalized, more abusive and less satisfying than I have ever known in my life.

Democracy has been reduced to a spectator sport. You sit at home and television tells you who is going to win the World Cup, who is going to win the Olympic Games, and who is going to win the next election. Real democracy is about what we do, where we live and where we work ourselves, and I think scientists could greatly help to raise the level of debate, because science starts with the facts. Scientific discipline is needed to understand the laws of nature and to understand the laws of society. The methods may differ but the application of knowledge can eradicate superstition, and it is the failure to bring knowledge and understanding to bear on some of these issues that I touched on that seems to me to be one of our greatest dangers. In the old days of course before scientists began work, when there was a flood or a fire, a witch doctor would say 'God is angry', and so someone would kill a chicken as a sacrifice and hold it up, and say that God would not be angry and there would not be another flood. Then when some bright lad came along and said if you dammed the river there would not be a flood anyway, superstition faded. Now we have new superstitions that if we get four million unemployed and hold them up, the IMF will forgive us and allow us to continue to proceed. But it is all witchcraft really, and we should now turn our attention to the realities of the waste of human resources and the denial of human rights.

The political and public debate in which scientists can quite safely participate should be at least as much about the objectives as about the methods, the machinery of government, and the management of the economy. Profit and loss is not a sufficient criterion for judging things. I have been trying to turn my mind to where the market will go next. It will not be long before elections are put out to private tender. Then on polling day the Returning Officer will say 'Mr Robert Maxwell has put in a bid for half a million for Chesterfield, Mr Benn has put in £50, so I declare Mr Robert Maxwell the new MP.' This is an attack on democracy, on the right of people to band themselves together. People want hospitals, houses, schools, to solve their problems. They are essentially political matters, because the allocation of scarce resources is essentially political. Scientists and engineers are citizens too. Although their skills may be different, their needs do not differ very much in any way from anybody else's.

If it is not about profit, is it perhaps about seeing everyone has a job, a home, education, good healthcare, dignity when they are old, peace, and can enjoy a sustainable environment? If those are the objectives, then you have to try and begin thinking about the methods that might be used to bring these things about. One of the most important methods at our disposal is that knowledge belongs to humanity. Secrecy is the greatest weapon at the disposal of those with power, and if things are opened up so that people can hear what the arguments are, before the decisions are taken, we might get somewhere.

We must get back to the idea of representation rather than just management. I read in the papers the other day that there are now more managers in Britain than there are industrial workers. There are lots of people in politics hovering about offering to manage you, but how many people in politics are representing you. I think there is a genuine crisis of representation. In the old days, to combat injustice, the Church would say to you 'Well, we know it's a horrible world, but if only the rich are kind and the poor are patient, it will be alright when we're dead', but people said they would like justice now! There is a new version of that: if only you keep your head down and do not rock the boat, and keep your options open, it will be alright when this or that party gets into power. But people want to have some opportunity to influence events now, and openness and representation are absolutely critical. Do not think I am saying this purely in a party context. We want to have decentralization as far as we can, because conditions differ so much—the historical conditions, the cultural traditions, the economic state of development. You cannot squeeze everything into one formula and apply it rigidly. My own experience of life is that very few changes are ever brought about by people at the top. By the time you get to the top it is very easy to forget what was the great popular pressure to put you there. Most change comes from underneath. If you put forward an idea, in the beginning it is totally ignored. Then if you go on, you are mad. Then if you persist, you are dangerous. Then there is a pause and you cannot find anyone who did not claim to have thought of it in the first place. That is how human rights were won.

At Shoreham and Brightlingsea, the police turned up with their riot shields to find they were striking old ladies in woolly hats. Have you ever known such attention given in the media to animal rights since people in their woolly hats went to Shoreham and Brightlingsea? That is why apartheid ended in South Africa. It was not because Nelson Mandela had a policy review or rewrote the Constitution of the African National Congress. It was because the Africans would not accept apartheid. Mandela was put in jail as a terrorist; he admitted it at his trial. Last time I saw him he had a Nobel Peace prize and was President of his country. If we had listened more, we might have understood better.

Progress depends critically on our understanding the meaning of democracy. In my life I have found five very interesting little litmus paper questions to ask powerful people: 'What power have you got?', 'Where did you get it from?', 'In whose interest do you exercise it?', 'To whom are you accountable?', 'How can we get rid of you?' If you cannot get rid of the people who govern you, they will not listen to you. Everyone in Chesterfield put me in Parliament, the bus drivers, the ticket collectors, the street sweepers, the doctors, the dentists, the miners. They are all my employers. But if I were the President of the European Commission, why should I bother to listen? The people cannot remove him or the head of the IMF. Scientists should study the basis on which progress is made.

My favourite quotation on leadership comes from a Chinese philosopher called Lao Tzu, who lived many hundreds of years before the birth of Christ. This is what he said about leadership: 'As to the best leaders, the people do not notice their existence; the next best the people praise; the next best the people fear; the next best the people hate. But when the best leader's work is done, the people say "We did it ourselves."' I found that so enormously encouraging, because the problems I have touched on are very difficult, and you are not going to find the answers coming from somebody on a white horse, or from a think tank. You are going to find them by trying to tap the experience of the human race and help them to realize that they have undiscovered capacity which could be deployed for that purpose.

TONY BENN

Born 1925, served in the RAFVR 1943–5, learning to fly in Africa and commissioned as a Pilot Officer. Served briefly in Egypt, and posted home at the end of the war in Europe. Transferred to the Fleet Air Arm as Sub-Lieutenant RNVR 1945, training at the Royal Naval College, Greenwich. Elected to the House of Commons sixteen times since 1950. Labour MP for Chesterfield since March 1984, and previously Labour MP for Bristol South East 1950–60 and 1963–83. Disqualified 1960 on the death of his father, Lord Stansgate, but re-elected 1963 after successfully fighting peerage law and renouncing title. Elected to the Labour Party's National Executive Committee 1959–93. Chairman of the Labour Party 1971–2. Cabinet Minister in every Labour Government since 1964. Candidate for the Deputy Leadership of the Labour Party 1971 and 1981, and for the Leadership 1976 and 1988. Holds four honorary doctorates from British and American universities. Author of: *The Regeneration of Britain; Speeches; Arguments for Socialism; Arguments for Democracy; Parliament, People and Power; The Sizewell Syndrome; Fighting Back*; a series of *Diaries*, and *Common Sense* (with Andrew Hood). Married to Caroline Benn, President of the Socialist Education Association.

Reproductive fallacies

JACK COHEN

Introduction

In this discourse I will consider Cinderella's shoes, the slyness of foxes,
why there are so many spermatozoa and the extent to which our mind-
representations are congruent with events in the non-mental universe.
Within the cat's cradle I hope to build, strung out in intellectual space by
attachments at these points and many others, I will concentrate our
attention on that part of the web which relates to our notions of reproduction and sexuality, and show that this volume has some intellectual
pathologies. Whether I represent these incongruencies as gaps in the
intellectual framework, or as strains in the web—or even, more biologically, as encapsulated insects whose wrappings hide corruption and
decay—their existence is very odd.

For most of us, even those of us deeply and professionally involved
in the reproductive field—as family-planning doctors, tropical-fish
breeders or agriculturists—this set of inadequacies is not available to
introspection; it requires some disciplined mental effort. I have found
them, I believe, because my professional involvement across, rather than
within, reproductive biology has thrown these incongruities into high
relief. You may, at the end of this discourse, feel only that I have aired
some minor biological errors of lay or professional understanding, in a
somewhat lubricious—perhaps even prurient—manner, and that these
need no special understanding beyond recognition of particular social
stances: Victorian, contemporary feminist, or rugby-club. But if I have
succeeded I will have shown you that, independently of this stance, this
intellectual area is notable for the *shape* of its gaps. The oddity is that we
have got so much simply wrong in what is arguably the most important
area of our social and biological lives. I hope to show an 'equality of the

gaps', that the gap between the original Cinderella fur slipper and the glass one is the *same* gap as between gynaecological realities and gynae-cological models, and that this is nearly the same gap as between Nature's versions of life-in-the-wild and the Disney versions. And that the causal simplicities of Hartsoeker's preformationist spermatozoon with the homunculus inside, which is echoed today by DNA preforma-tionism, are separated by a similar gap from the real developmental causalities of frog embryo or human child.

Finally, I will guess as to the social and emotional insecurities which the gaps protect from 'the real world'; why do we get the science so wrong? This is a large task, and I hope to increase its difficulty by expla-nation, within the scope of one paper, rather than by assertion; I will fail in this to some extent, and will point towards other sources in this diffi-cult area.

The reproductive arithmetic of wildlife

An easy place to start is with some reproductive arithmetic. The issue is clear, obvious—and contrary to our prejudices in an entirely public way. For my argument it serves as the thin edge of a didactic wedge whose hammered back is our ragged understanding of our sexuality. We compare the romance of animal life in the wild with animals subjected to domestication, with 'factory' farming, with laboratory exercises. Such comparisons have always portrayed the wild animal as 'proper', free, exercising its abilities to the full in a long and usually exciting life. By implication if not assertion, it is concluded that most animals in the wild fulfil their biological natures, whereas those in captivity have lesser exis-tences along a variety of axes.

There is a simple way in which this romantic belief is the reverse of the truth. A breeding female starling, in the Home Counties of England, lays (on average) about 16 eggs in her reproductive life. The number of starlings is not going up much or down much, so *two parents in this gen-eration only produce two parents in the next*; arithmetic insists that 14 starlings die before breeding for each two that breed [1]. Ecological observation, especially of food chains, shows us that most of these 14 die as nestlings or just-fledged young birds. Similarly, of the 10 000 eggs laid during her lifetime by a female frog, or the 40 000 000 laid by a female cod, only two on average breed: 9998 or 39 999 998 die, very nearly all as babies. The vast majority are eaten by other creatures, die of suffocation complicated by digestion, or have less parts than required for continued life. Some are not eaten directly, but decay; some ponds dry out, and the tadpoles split as their skins dry, and some of the baby cod die because of excessive parasite load (*all* wild animals have *some* parasites), but such cases are rare. Young animals are nearly always eaten. They do not, as Disney assures as—and as we would much rather believe—mostly grow up to have fulfilled adult lives, mating and begetting, dying romantically in the accomplishment of some great task. Nature is not a good pet-keeper, does not work by Home Office rules as laboratory scientists— and their experimental animals—must. Captive animals, indeed, very nearly always have a much healthier, longer life—and a much shorter, less painful death—than their relatives in the wild. And I include de-beaked chickens (now mostly historical) and bulldogs, laboratory rats and even de-clawed cats living in urban apartments. Aquarium stocks of blue gouramies (*Trichogaster trichopterus*) provide us with a more subtle example. People buy them at about 3–4 cm; a few breed at about 7–8 cm, and none reach 15 cm, but die after a calm, boring life of a couple of years in a home aquarium. Amazingly, the same species is the national fish of Singapore: nearly all die—are eaten—under 2 cm but a tiny proportion breed at 40 cm and die at 50 cm or larger, occasionally living for ten years. Which is 'better'? We hate thinking about death, but we do not fund Samaritan stations for lemmings or baby cod. Some of us do, however, release captive mink to a painful, short parasite-ridden existence in the wild and a shivering, starving death in the name of animal liberation; the few which survive generate misery for otters, stoats and other indigenous mustelids. This is nearly as silly as putting aquarium gouramies back in Singapore, and much crueller. I am not arguing that we *should* be cruel to animals, just as I would not argue that the existence of so many traffic deaths makes murder morally OK. I am attempting to give a factual, rather than a romantic and *mistaken*, background to our decisions about 'natural' animal lives. Very nearly all sexually produced organisms in the wild have lives which are nasty, brutish and *very* short, and contribute to food chains. This, rather than the romantic view, should underpin our decisions about domestication, conservation and education.

There is a great gap between this reality and our romantic ideas of 'life in the wild'. I interpret this as part of a retreat from overt death, a genuine thanatophobia in civilized societies, which relegates most human death to hospital, animal deaths to discreet slaughter-houses, and whose honest symptom is vegetarianism. Related perversions are Elm Street and vampires, supposedly anti-agathic cosmetic substances and procedures, retreat from dissection in schools and the excesses of many so-called 'animal liberation' movements.

Nursery animals, make-a-human-being kits, nursery tales and sexuality

My second example concerns the way in which we teach sexual anatomy to schoolchildren. But in order to make my point, to take your minds to new places and make new comparisons, I need to range much more broadly. Indeed we must consider the whole question of the reproduction of human culture, including knowledge about sexuality, from generation to generation.

I have called this the 'Make-a-Human-Being Kit' [2], different in each (part of each) culture; it is usually recognizable as a succession of ritually constrained environments, as fixed as the succession of clothing or of stories. The progression of babies in Western societies from nappies (diapers) through blue jeans to adult clothing is matched by a succession of play-group/junior-school/senior-school/further-education environments— and by a succession of stories, from the simplest nursery-songs and tales to later classics of world literature. To a large extent these have been replaced by a Disney/Star-Trek/horror-videos sequence as television has supplanted reading, and this has worried those of us who believed that the sequence *we* grew up with was somehow a universally *right* one. Such replacements are legion in human history—indeed they may *be* the necessary, other-than-contingent, elements of human history.

Some of the deep similarities across cultures and through history are very impressive, particularly our commitments to animal images for setting up our emotional understanding, so beautifully exposed by Shepard [3]. Nursery tales of all human cultures use animal icons, even those most removed from hunting or agriculture; ours is mostly industrial but, significantly, we do still retain zoos in our cities, cats, dogs and budgerigars in our homes—and Teddy Bears and rocking-horses in our nurseries. We all hear stories of foxes, chickens, little pigs, wolves, bears and snakes; the *Jungle Book* pulls a particular set of these together, and is clearly a nearly lineal descendent of *Aesop's Fables*. The *Tales of Bilbai* (Pilpai) and the *Book of Beasts* [4], some of the *1001 Nights* tales and much of heraldry document this lineage.

For those of us in the British cultures, foxes are 'sly' or 'cunning', owls are 'wise', snakes are 'deceitful', and so on. A moment's reflection will show the absurdity of these attributes (when were you last deceived by a snake?), and comparative study soon shows their mutability: for the Inuit child, the fox is 'brave' and 'fast', the hero of the stories as he is the 'cunning' villain in ours. Neither is, of course, the fox's own fox, or the biological fox. They are icons for the character traits, symbols for our mind-building; it is probably more useful to think of 'cunning' (in all *our* minds) as being 'what the fox does in *our* stories' rather than thinking of the fox as 'being' cunning—and similarly for all the rest. Shepard has shown us that our establishment of all this important understanding in our children—in us as we grew up too—depends on our use of animals-as-symbols, so that foxy, chickening-out, mouse-timid, weasel-minded are all phrases which rely on this early programming. The reproduction of these iconic animals is very odd, incidentally. Eggs, nests and babies are given values which strongly suggest that we do not get our sexual/reproductive information in this way. In the *Muppet Christmas Carol* film, for example, the progeny of Kermit (a frog) and Miss Piggy (now Mrs Kermit) are two little Piggies (female) and a little Kermit.

The fairy tales we hear later, in contrast, are replete with princes/princesses/marriages, nasty stepmothers and spoiled children who get punished in cruel and unusual ways. Perhaps we get our sexual models from these icons? Certainly many jokes, especially perhaps about some New York Jewish culture, draw heavily on the 'princess' attributes. However—and now we draw closer to the point I wish to make—the stories we use now were very heavily edited, bowdlerized, standardized by such collectors as the Perraults and the Grimm brothers. The Opies have collected some earlier versions [5], and they were much more violent and much more overtly sexual too. One difference will suffice here, but there are many others which seem to follow the same pattern: the earlier Cinderella stories did not have a 'glass' slipper, but a 'fur slipper', clearly a euphemism (the suggestion in a German early-folk-tales collection [6] is that the Prince tried on all the girls' 'fur slippers', which is pretty overt). If this really is a part of the Make-a-Human-Being kit then our kind of people, post-Grimm, must be made differently from our pre-Grimm ancestors. Similarly, those of us, boys and girls, who have grown up in the last twenty years have been surrounded by women's bodies, in explicit (whether exquisite or disgusting) detail, on the top shelves of any newsagent, video shop or later evening television saga. Surely the boys and girls who hear of Cinderella's warm fur slipper become differently sexual people from those who hear of the cold, clean shiny glass slipper ... who are different, again, from those of us who have been surrounded by photographic images representing male (advertisers?) fantasies of female sexuality while we were growing up.

The make-a-human-being kit and the gynaecological glass slipper

Does it matter which tools, which icons, are used to build our minds? The previous argument, that sexual attitudes are subject to fashions in

"Oh hell! Are you sure? I was hoping we were lust."

Fig. 1 We use animals to exemplify human attributes (cartoon by Bud Handelsman).

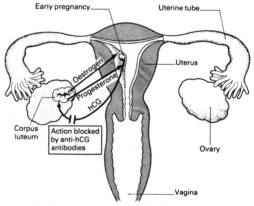

The immune method: anti-hCG antibodies stop early pregnancy
from maintaining corpus luteum

Fig. 2 An illustration of the female genital anatomy from a very good modern textbook on contraception by a gynaecologist. Note that it has cavities where none exist in 'the warm reality' (from *The Pill* by John Guillebaud, Oxford University Press, 1993).

the nursery, should perhaps be checked against the Inuit child who thinks foxes are wonderful, while we think they are rather nasty. There are obviously many different Make-a-Human-Being sets, which work in—and construct—different cultures. We do not often meet foxes, and our pre-judgement of them is not perhaps very important. We do enjoy, and think we understand, Shakespeare or Chekhov even though our shared iconic background is very different from their intended audiences. Equally, our courtship rituals, and more general relations between the sexes, have a long history of modification from the medieval chivalry/Courts of Love/troubadour traditions, which were overlaid onto a paternalistic peasant/feudal tradition, and which have cycled around Elizabethan and Victorian stances. Our current romantic notions, then, derive from a mixture of homosexual and indigent-to-rich-patroness attitudes, expressed in troubadour verse and literature, which were transmitted mostly by the clerics of an anti-sexual and anti-feminist Church. We build our present sexual attitudes on this very mixed historical tradition, and try to create our forbidden/disgusting/embarrassing/allowed code of behaviour, and to transmit it to our children. In our culture, we permit the media to bombard us, and growing children, with the greatest variety of sexual/romantic icons the world has ever known, which makes the background 'noise level' impossibly high.

We are each sure, nevertheless, that our own special blend of feelings about sexuality is The Right One; David Lodge [7] has convincingly shown, however, how rapidly we move between such Right positions in response to cultural flows. While overtly valuing multiculturalism and the reproduction of subcultures within it, we encourage the media, especially the advertising media, to produce and disseminate sexual icons whose success depends on their dissonance with accepted social practice. While we can accept this plasticity for our behaviour and attitudes, we generally do not realize that such substitution of icons for reality extends even into our vision of our sexual anatomy.

Most of us have not been embarrassed by meeting foxes when in the company of Inuits. But men and women, boys and girls do meet female genital tracts in a number of personal and social contexts, and it seems to all of us important that we share something like the same set of rules for handling sexuality socially. I am going to suggest that, like the use of the fox to symbolize slyness or the owl to symbolize wisdom, our set of reproductive/sexual icons is a peculiar one. It is related more to emotional mind-building than to anatomical science. An educationalist might say it was 'incoherent': the teaching and the subject are in different universes of discourse. The standard school picture of the human female tract, indeed the standard picture in the family-planning textbooks too, shows an interestingly diagrammatic vase-like structure with a wide cavity in the vagina and an equivalently 'empty' uterus. Asking girls, or indeed women, how they locate that diagram in relation to themselves achieves very interesting answers to questions like 'how big?', 'where in relation to orifices?' and so on. Most that I have asked have related it to the diagram in the sanitary-towel packet, not either

diagram to the warm reality but diagrams to each other. We know the personal reality of our own body parts as finger-senses, mostly not as optical pictures. Diagrams of vulvas, even photographs, are very difficult for us to relate to real body-parts (photographs of penises are easier, probably because the organ is so much more overt).

We all drew the female-genital-tract diagram, vase-shaped with its open gaping vagina, at school or wherever, and we (mostly) could have known it was wrong as a representation. It is in fact a diagram of a for-malin-fixed, removed and sectioned organ, which has shrunk to create internal cavities—there are many such in the Hunterian Museum in Lincoln's Inn and in most Medical School Museums. Medical students have been taught from such grossly distorted specimens for hundreds of years, but there seems no reason why we should go on using this out-dated formality (*sic*).

Is there a 'better' reality from which to draw our models? We might see the uterus from its outside, its peritoneal surfaces. There are now many video-films from the days when human eggs were re-covered for fertilization *in vitro* (IVF) by laparoscopy, showing the

Fig. 3 Vesalius' drawing of a removed, opened-up vagina and uterus—perhaps the origin of the strange iconic usage of Figs 2, 4 and 5. Vesalius and his pupils, hearing of her death, snatched the body from the tomb, but, unfortunately, the monk together with the parents of the girl complained of the outrage to the city magistrates so that the anatomist and his students were compelled to dismember and free the body from all skin as rapidly as possible to prevent its being recognized. Since they had stolen the body expressly to examine the female organs, the best they could do was to encircle the external genitalia with a knife, split the symphysis and excise the vagina and uterus in one piece after severance of the urethra. Later, the uterine cavity was exposed by sectioning the body longitudinally and turning up the anterior half. We imagine that the uterine tubes were lost in the hurried method of preparation.

In the Vesalian terminology the uterus consisted of the fundus or body, the uterus proper, the cervix or neck, which is not the cervix of the modern anatomist but the vagina, and the vulva. The illustration therefore shows: A,A,B,B, the uterine cavity; C,D, a slightly elevated line compared to the raphe of the scrotum; E,E, the muscular wall called the internal or proper tunic of the uterus; F,F, a protuberance of the wall into the uterine cavity; G,G, the cervical canal; H,H, the external or peritoneal tunic; I,I, the broad ligament incorrectly shown extending along the lateral border of the vagina; K,K, the divided edge of the vagina; L, the urethra, in Vesalian terminology the neck of the bladder. The rest, referring to the vulva surmounted by the pubic hairs, is self-evident, says Vesalius. Note the cut edge of the skin encircling the vulva and indicating the method of the urgent excision.

vagina–cervix–uterus, the oviducts and ovaries and the other viscera in a gas-inflated peritoneal cavity. But why *should* we wish to see their outsides? Does it really help? Most people do not want to think of their sex organs as visceral pictures, which still produce disgust. Perhaps, as many educationally wise family-planning doctors and biology teachers believe, the functions are indeed best explained by the classical diagram, the vase icon, especially as it is indeed the one they will see in all but a very few books. (Many modern books now show a vertically sectioned woman, with parts in place but difficult to distinguish—easier for learning to place tampon or diaphragm, but very difficult to relate to the school picture.) However, I do not want to ask how we *should* teach the subject; I want only to emphasize the distance between the warm, living reality and the diagram we use as an icon to teach about it.

Particularly, I wish to draw attention to the models used by doctors for instruction of other medically qualified people how to fit intra-

How the human egg is often deceived.

Fig. 4 Larson's famous cartoon about fertilization; it makes fun of the icon, as so many of his biological cartoons do. The Far Side © Farworks, Inc./Dist. by Universal Press Syndicate. Reprinted with permission. All rights reserved.

uterine contraceptive devices ('coils' or IUDs, better IUCDs). These *very rarely represent the reality*, they model the icon. This, surely, is like believing that foxes are not only sly, but also dressed up in hunting waistcoats and twirling their whiskers with their paws! Perhaps it is too embarrassing to instruct the plastic-model manufacturers: 'Use a nice soft-but-elastic material—moisten it and try it with your fingers—and look at this video for the outside shape . . . you know what the bottom end should look like! Now make us a model which is like the personal bits of women.' (Perhaps the makers of 'sex dolls' *have* responded to this challenge?) What we actually have, for doctors to practise with, is a variety of devices mostly in hard, Perspex-like materials, which precisely represent the icon, not the reality. They are gynaecological glass slippers.

Even more absurdly, IUCD *designers* obviously use them too; this is, surely, like the designers of real wolf-traps putting in the address of the Three Little Piggies as bait! Perhaps we should not be surprised that the Dalkon Shield, clearly designed not to fall out of the kind of space shown in the icon—but absent in the real warm uterus—resulted in so many septic abortions. It killed hundreds of women and made thousands infertile; the string was blamed, but the shape is entirely inappropriate too. It was iconic, not biological, and the dissonance was lethal.

(A note should be added here. The IUCD *should*, of course, produce a pathological state: infertility. It is clearly required that it should be biologically inappropriate for uterine functioning. But the various T- and 7-shapes with copper wire, hormone-release plastics and other optional extras conform to the real uterus, more or less; and the older spirals specifically did *make* a cavity, and so prevented implantation. The Dalkon was particularly suited to make my point because, *although it may not have been possible to predict that it would*, it did cause much extreme pathology.)

Another example of the way medical scientists and doctors have been persuaded that the female-tract icon represents reality concerns the replacement of early human embryos, 'test-tube babies', into the uterus. While Edwards and Steptoe were achieving successful fertilizations in the early 1970s, with apparently normal early development *in vitro*, they could not get pregnancies until 1978 (Louise Brown was the first birth). We embryologists knew that we could not guarantee normality of this early development simply because it looked normal down the micro-scope, so Edwards and Steptoe were inclined to blame culture condi-tions (hamster eggs *in vitro* look as if they are developing normally, yet pregnancies are sporadic even with modern laboratory culture tech-niques). However, they were putting the embryos into the uterus in about 0.1 ml of medium, assuming (as we all did—from the icon) that there was about 1–3 ml real fluid space in the human uterus. That is wrong; the walls are close together, and the notional volume is about 10 µl, or 0.01 ml, a tenth of the volume they were putting in with the egg. It is probable, from present experience with small volumes, that these early eggs flowed out with the excess fluid [8]. We now know that the early embryo in the uterine 'cavity' is more like a ping-pong ball under a rug than like a goldfish in a bowl of water.

Equally, this mistaken picture gave us strange ideas about sperms. The usual metaphor, of salmon swimming upstream to breed, is misleading in very many ways—but is sufficiently pervasive that everyone under-stands the Woody Allen *All You Ever Wanted to Know about Sex* ... sketch. The idea of sperms competing with each other in a kind of obs-tacle race in a swimming pool is, for some reason, very attractive—as an icon. The simple model has been replaced, for most biologists [9], with between-males competition models—with optional 'kamikaze' and 'female-choice' extras in the later, more sophisticated versions of the computer game [10]. But it still cannot accommodate many observations, such as the obvious coagulation of human semen which only releases the spermatozoa 15–20 minutes after ejaculation [11], or the many experimental data which demonstrate that it is a special group, a very tiny number of spermatozoa which is permitted to the vicinity of the egg to fertilize [12].

Now that embryologists in many infertility laboratories are picking out random individual sperms to fertilize eggs (in cases where the man's sperms are immotile, or too sparse, for regular IVF which leaves some female selection in place), we should recognize that human fertilizations of a wholly new kind may be occurring—sperms may be fertilizing which never would have been permitted through by the female tract. If we retain the iconic picture, the image of a coy egg with invading-barbarian sperms racing for the prize—all of them capable of rape if they get there—this worrying possibility cannot arise. Only if we have a more science-based model can we articulate the problem. (However, it must be said that even those of us who believe that it *is* a special tiny population of different spermatozoa, which the female tract allows through, do not anticipate congenital abnormalities in these embryologist-selected progeny.)

This possibility of differing states of alarm about laboratory fertilizations emphasizes that all scientific questions, predictions and theories are dependent on metaphors and models, images and icons. When we rearrange the details of human fertilization in the laboratory, specifically to be independent of sixty million years of evolution of the sperm transport system, we should be very sure that our models are useful, creating new human beings safely. Particularly, when we know for sure that our iconic models of sperm transport cannot 'work', we should be more anxious to construct new models which are less dependent on socially constructed anatomical icons. Only then can we judge the safety of new IVF procedures. As the economists reputedly say 'It works fine in practice; now we have to see if we can get it to work in theory.'

DNA replication and authentic reproduction

We have seen that both the arithmetic and the anatomy of important reproductive events have a large conceptual gap between their popular icons and their realities, and I have symbolized this gap by contrasting the gynaecological glass slipper with the living, warm 'fur slipper'. DNA is the most popular reproductive icon today, and my last examples concern more gaps between real reproduction and the DNA-linked popular images.

Many famous biologists have found reproduction very puzzling. Darwin,

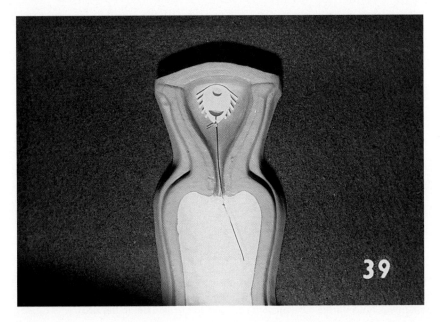

Plate 31 A picture, used for teaching doctors how to remove IUDs, which uses the same convention—not like the 'warm reality' at all (from Camera Talks).

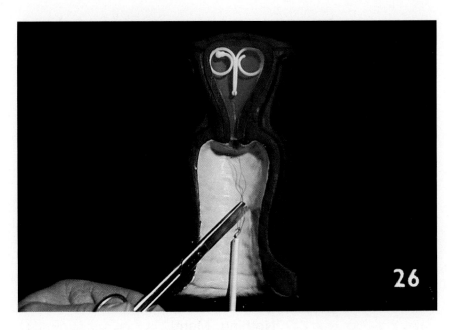

Plate 32 Another slide from the same series as above, showing the Dalkon Shield 'in place'; it is clearly designed not to fall out of the icon. It caused much pathology when used in real women (from Camera Talks).

Plate 33 The Belousov–Zabotinski reaction: a recursive chemical reaction, demonstrating that complexity can arise from simplicity, that the complexity of a human being can result from interactions during development and need not be laid out in the DNA.

"Nothing, Mom!"

Plate 34 DNA cannot 'make' organisms (Reprinted courtesy of *OMNI Magazine*, copyright 1988).

according to the biology textbooks, believed in 'blending inheritance': the offspring had intermediate characters between those of the parents. However, such a belief could not fit with his natural selection theory, because any specially good (or bad) character would be diluted out in subsequent generations as organisms carrying it bred with the general population, which did not carry it. Darwin must, however, have been surrounded by people who said 'Hasn't the little darling got the Darwin ears, then?' or 'Look, he's got his mother's nose.' Especially among the Darwins and Wedgwoods, there were many such characters that contradicted 'blending inheritance' very obviously. So Darwin could not easily combine his model with either his theory or his folk examples. However, a moment's reflection shows us that 'having his mother's nose' does not answer any questions either. Clearly the nose, or indeed the nose's shape, does not pass across the generations.

Two hundred years earlier, it had been fashionable to believe that it did, in miniature. Those who believed, like Harvey, that within the egg there was a complete embryo-ready-to-make-an-adult, also had to believe that within those eggs, too, there were still smaller adults with yet tinier eggs. .... Hartsoeker and others were sure, in contrast, that each spermatozoon had a 'homunculus' inside, and again this preformationist model required that the homunculus had sperms within, and they in turn ... and so on. Chinese boxes, or Russian dolls, are easy in comparison (because the next-one-down *fills* its 'mother'). Chemical ratios, and hence Dalton's atoms, showed that this continuous sub-division could not go on for many steps, so that preformationism was a false model (except, perhaps, for those who believed in a Day of Judgement, when all the creatures were sterile because their babies would have been smaller than atoms at the Creation!). Apart from the difficulty of 'Was it the egg or the sperm which carried the characters?', both Ovists and Animalculists lost credibility as thoroughly as those who believed that phlogiston accounted for combustion.

More than a century after Darwin, preformationism has now come back in a new guise: the nose-shape apparently *does* pass across the generations. It is there, we are told, in the DNA information in the chromosomes, only requiring to be read out during development to give baby Sarah her mother's nose. This, I am afraid, is another case where we mistake the icon for the reality. But because the promoters of this view have even more scientific credibility than Harvey (who only discovered the circulation of the blood), the demonstration of the poverty of DNA-preformationism will require that we creep up from unexpected angles on the problem.

Ian Stewart and I have dealt with this problem in several papers and books [13, 14]. The basic stance that we are criticizing is exemplified by Rucker: 'Your hand is designed according to certain instructions coded up in your DNA. The length of these instructions gives a measure of the amount of information in your hand' [15], or all the various claims that the Human Genome project will reveal, at last, the essential nature of Man (Woman is usually not included, perhaps because she is recognized to be more subtle). Its identifier is the use of the word 'blueprint', comparing the 'DNA instructions' with the blueprint for a car or a Walkman.

There are four steps to my argument that DNA is not a blueprint. First, a real industrial blueprint is not a 'blueprint' in the iconic sense: even the car or Walkman blueprint is not anything like *complete* instructions for building the item. It assumes the existence of lots of outside structure, lots of information in its surrounding context. When it specifies 'capacitor Z122, 0.001 mf' (or item '8A40Z122001'), or 'put bolt/nut 32 into holes 13 and 28 and tighten' there is a lot of structure, information, in the outside world it is relying on. Most of it is in the mind of Fred who is doing the assembly, and who knows that bolt 32 is 8 mm and needs *this* pair of spanners; Mary knows that all the 8A40 components are kept in the drawer under the clock and what to do if the stock runs low (but not, usually, how capacitors work). There is usually a picture, on the real blueprint, of how the finished article should look—and some figures for how it should behave in certain circumstances. The DNA is both more and less than this. It is more in that it specifies Mary and Fred, in the sense that the cell organelles which translate DNA code into proteins, ribosomes, are made of proteins and RNA which are themselves specified in the DNA. But it is less in that it does not have a picture of the final product in any sense. That is why the thalidomide tragedy should be blamed on the 'blueprint' approach. We all knew in the 1960s that the baby, two arms and two legs, was specified in the DNA ... and we did not think that the drawer under the clock had perhaps been filled by quite the wrong components. That kind of thing cannot happen with real manufacturers' blueprints, because there is a quality test at the end. In a subtle, Darwinian sense there is quality control for organisms too—but as we have seen above, nearly all of them fail!

The second point about DNA is that it does not correspond to the chemistry icons we acquired at school, it is not 'just a chemical'! The usual, iconic use of 'chemicals', indeed, refers to nasty, simple crystalline solids or smelly oily liquids that we call E-numbers and put into food—they are *artificial*. The idea that the food, too, is entirely made of chemicals does not occur to us (though we recognize its truth, of course), and we are encouraged in our belief that chemicals are artificial, hormones and vitamins are natural by the use of the adjective 'magic'; the magic Vitamin C molecule which helps you not to get colds, the magic antibiotic molecule which kills bacteria, and especially the magic DNA molecule which ... self-replicates, it says here [16]. It does not [17]. DNA in a test-tube, even with all the components of more DNA, just sits there; it needs a vast and complicated machinery—much more complicated than DNA—to replicate it. Nor does it make animals, plants, even bacteria. It just sits there. If DNA is not so special, so magic, what is?

It is the complicated machinery, chemical machinery to be sure, of which the DNA is a functioning part which does the work; the parts of the machine are specified by bits of the DNA so that *in the context of cell machinery* more cell tools are made. But the DNA does not make them, or replicate itself. This machinery is not very remarkable, chemically; some of our more modern industrial chemistry is nearly as complicated, with many catalysts just as the cell has enzymes, and many processes renovating those catalysts and recycling energy and materials. DNA is *not* such a remarkable chemical at all. But our chemical ideas come from school chemistry, which was committed to the concepts of 'Equilibrium state' and 'Terminal reactions'. That is to say, 'What has happened when nothing more is happening?', when the precipitation has finished or the pH has come to its final value. Biologists use words like equilibrium, terminal, final to mean 'dead'. So it is not very surprising that such chemistry sees biochemistry as 'magic'. But in the real world of rusting cars, translocating clays, forming raindrops or eroding mountains nearly all the chemistry is cyclic, periodic, recursional—it uses products of reactions to autocatalyse reactions just like biochemistry does—and as school chemistry does not. The Belousov–Zhabotinsky reaction [14, 18] is just such an *ordinary* reaction. But it has taken forty years to persuade school-thinking chemists that it does not upset any 'laws' to have cyclic chemistry, and that if the Second Law of Thermodynamics apparently does not permit something which can happen, so much the worse for the 'Second Law' [14]. The origin of life, it turns out, was not so miraculous, so 'magical'. Clays, rusting iron, the organic chemistry of meteorites are all on the path to life—but school chemistry icons are not. And DNA is not magical.

The third step is best explained by the use of a tape/tape player analogy. Many attempts to see DNA as the 'information content' of an organism have likened its linear array of four bases to words on a page or to magnetic signals—basically N or S—on a recording tape. Let us imagine that we find such a tape, but on Mars; there are no Martians left to tell us how to play it. Can we tell what is recorded? We find that the open end of the tape (which may be the beginning,

the end, or instructions for the player to do something ...) has
NNNNNNNNSNNNNNNNNSNNNNNNNNS for 64 repetitions. Further
along, the tape has a sequence with no apparent pattern—but nearly the
same sequence is found a little further on. From this evidence, we cannot tell whether the tape is a sound tape, a speech or music tape, a video
tape, program for a computer or instructions for an automatic machine
like a lathe. By the repetitive patterns we might decide whether it were a
digital or an analogue tape, or if a video tape what its frame size was. But
I suspect that what we would do is build a variety of players and try out
(copies of) the tape. Different players have different requirements from
their input tapes: at one end is the nickelodeon, which requires just two
digits and a coin to produce any of 50 or more pieces of music; at the
other is the computer program, within which each and all of the signals
could be necessary to its function. If we offered a prize for the first team
to 'decipher' the tape, it would not surprise me to find that the video-
player team soon gets a picture of what looks something like a Martian
Marilyn Monroe undressing, while the audio team achieves some very
interesting modern-sounding music, the engineers at the automatic lathe
produce a very creditable sprogget—and the computer-game lot will
have found a way to use it to set up a Martian Quest adventure game. In
all the different contexts the message would 'mean' different things [14,
19]. Perhaps, however, the tape was *really* the DNA-equivalent of a
Martian life-form. In that case, *if it is like Terran eukaryote organisms
which develop*, we would need the right Martian egg to put it into,
because—as we have seen—the character of the tape-player is at least as
important as what is on the tape.

That leads to my fourth and final point, about the gap between our
informational icons about DNA and what it really does. This concerns
the relationship of DNA to the development it supposedly controls—for
the shape of higher organisms is attained via development. This is not so
with prokaryotes (as a rule): bacteria simply grow by making more
molecules and fitting them into the structure, then dividing when a certain size or state is reached (using different genes, different molecules
and controls), then growing again as daughter bacteria. Most eukaryotes,
in contrast (even non-cellular eukaryotes like *Paramecium*), have a process of development: the mature, breeding animal has a different structure, not only a different size, from the growing stages [1]. The human
egg, for example, becomes a blastocyst which in turn makes embryo and
placenta; the embryo then uses maternally supplied material to make
itself into a fetus, which becomes a baby when it is born. Successful
babies become children which, subject to different Make-a-Human-Being
Kits, become pubertal; only then do they produce the biological equip-

ment which can generate human eggs and blastocysts, but rather more development is needed before they can actually reproduce more (cultured) human beings. Frog eggs cleave, making little blastulae which gastrulate, preparing the embryonic geometry for making the nervous system in the 'phylotypic stage' or neurula, which resembles that of all vertebrates including us [17]. Then it develops gills, sense organs, kidneys and a working gut and hatches as a tadpole. A few tadpoles grow up to metamorphose into little frogs, and a fraction of those get through their first couple of years and make sperms or eggs; then some breed [1].

Up to the phylotypic stage, development is not 'controlled' by the egg's DNA at all—many experiments have shown that the early nuclei are simply 'passengers': if they are prevented from transcribing their DNA, or even if they are replaced by nuclei of a different species, development follows the original path to the egg's phylotypic stage [1, 17]. The egg's architecture, and many special informational and other molecules acquired in the mother's ovary, control early development. They therefore control which phylotypic stage it becomes, a worm embryo or a vertebrate embryo—or something else, perhaps non-viable. Because the nuclei are now in different parts of the embryo, they can be turned on differently by the different cytoplasms. Only when the nuclei are established in cytoplasm of different properties, calling for different spectra of gene activation, do the different cells of gut, nervous system, kidney and skin diverge as they use their nuclear information.

A rather simple-minded but useful view of this post-phylotypic development imagines each cell as having a 'map' and a 'book': the cell finds out where it is in the embryo, by local cues, then looks in its nuclear DNA book to see what to do. This model of development [20] emphasizes that the map is drawn from the mother's DNA book, so is happy with the idea that the structure of the later embryo, and the adult, is referable to the information in the DNA. There are many experiments, too, which show that the map is much the same for rather different embryos, and indeed that the book has much the same chapters too (gene families are common to very many different kinds of organisms). Why then are organisms within the same phylum as different from each other as frogs, fishes, rats and human beings (all vertebrates) or snails, oysters and octopuses (all molluscs), while the very similar organisms of one species usually have some 10 per cent of their gene sequences different?

It must be that some small differences in DNA sequence have major effects on development, while some major differences have little effect. Perhaps the crucial differences are in the genes whose products act through the ovary and egg to affect the phylotypic stage, whose details radically affect all later development. And if that is so, then the whole idea of mapping the organism's final shape from the DNA sequence falls to pieces: if small changes in one part of the program can have effects which change the action of lots of other bits of the program, then there can be no one-to-one mapping from genes (sequences of DNA) to characters like a hand or a wing [13]. Indeed, it would be expected that very similar organs, in two different descendants, will have different developmental programs—and each different from the ancestor [1]. (Metanephric kidney in birds and mammals, or the inner ears of different vertebrates, are good examples; the argument here is that *if* the kidney program in the bird were exactly the same as in the mammal, the kidneys would be very different because the embryos are so different—in order to be so similar the programs must be different!) When comparing *different* groups of organisms, especially those with different phylotypic stages, we should think 'nickelodeon' (in which very different tunes can be played as a result of very small differences in tape input; the relationship between typing in 42, and getting the Hitchhiker's Guide music, and typing 41 and getting Beethoven's Fifth, is an entirely arbitrary contingent). Within a species, however (in which these major developmental events are all set up in the same way), developmental control might be more like tape for a cassette player, where each copy has slightly different magnetic sequence and yet produces effectively the same sounds. In the latter case there *could* be a one-for-one correspondence between parts of the tape and particular musical phrases from the speaker (but only for the one piece of music)—and *differences* could usually be attributable to differences in the magnetic sequence. In this analogy, notes would be like individual proteins, perhaps, while phrases would be characters—different individuals of a species could be like different performances of the same piece of music, about 10 per cent different but not in important ways.

I have previously used 'thought experiments', rather than the above argument, to show that genome does not map to phenome, but that differences in the same species do map. I have used a 'worm and a fish' [2], or 'hypnoceros and frogodile' [14], supposed to have exactly the same DNA but used differently and repeatedly in successive generations, to show that mapping from the assembly of genes to the assembly of characters just will not hold up as a concept. Of course, *a difference* between genes can and does account very usually for *a difference* between characters—if you have normal pigment and I am an albino, this is because our DNA tyrosinase sequences are different. If you have blue eyes and mine are brown, if we have different blood groups—even if you have too many or too few fingers and toes—that can be referred to particular places where the DNA is different. But this does *not* mean that there is information in the DNA which 'makes a hand' or a foot, or an eye, or an altruistic act, or a person [13].

Why have we so exaggerated the creative, descriptive power of DNA? Why are we so tempted to see it as description rather than as prescription, as blueprint rather than knitting pattern? Why do we cling to the icon, DNA-as-photograph, rather than the reality, DNA as one of the complex inputs which together predictably interact to make our babies? I think the answer lies in our reproductive insecurities, just as it does for Disney wildlife icons and vase-shaped uterine icons.

Conclusion

I have blamed our thanatophobia (fear of death) for our wildlife arithmetic errors, but I do not think it is as simple as that. Our less-violent post-Grimm nursery stories have fed back into the Make-a-Human-Being Kits of all Western cultures, and there is a complicity between the adult icons and the nursery stories which is not stable, but evolving with its own dynamic [2]. In our cultures we have lost any balance or equilibrium such as *must*, almost by definition, have been found in ancient, unchanging cultures; innovations as various as baby-milk, aeroplane flight, vitamin tablets, acceptance of one-parent 'families', stainless steel, polythene bags have arisen, and fed back into the culture. Which advertisements, or novel plots, move us and change our minds must in turn affect how we expose the next generation to such influences, what films we make, how we judge child-abusers or heroic criminals. This is not controllable, because it is a recursive complicity [14], like the development of an organism; there is no reason to expect cycles, in this view. Anything can happen, and our prophet is Alvin Toffler [21].

Within such a changing culture it is tempting to explain all our oddities by incoherence: there is no way to tell what effect any change will have, so even the most puzzling and consistent ones, like the reproductive misunderstandings above, are just examples of this incoherence. However, Toffler managed to predict the fall of the Iron Curtain countries about ten years before it happened, and even to predict that 'grey' politicians would supplant the 'heroic' figures in the 1990s [20]. Like Toffler, I believe that we can find an internal coherence in the set of changes, even though individual oddities may not be referable to specific causes.

A major part of our new, innovative culture is a strong concern with our children's inheritance, both biological and cultural. Such concern is, of course, only appropriate in a changing culture; in a stable, reproducing culture it is simply accepted that children will conform to tradition. (The musical *Fiddler on the Roof* puts Sholem Aleichem's descriptions of these issues in ghetto culture into more generally accessible language.) This concern is different from that which the post-mature generation have always had, from Roman senators through ghetto rabbis to Conservative politicians, that the next generation has not been brought up properly and will take the world to hell in a hand basket. It relates more to the very well-founded supposition that many of our children have been conceived from 'extra-pair matings' [10], as has been revealed for so many apparently monogamous animals by DNA-fingerprinting of the offspring [9]. With a good case having been made that nearly 1 in 5 of us is of other-than-assumed paternity, the natural anxieties which men have about their paternity have been justified. This paternity information has not been generally available [10], but it is entirely consonant with the increased permissiveness, the covertly and overtly advertised high incidence of adultery, together with societal approval of such symbols as red sports cars. The very highly charged debates about the extent to which our character and abilities (symbolized by IQ, for example) are contributed to by 'nature' or 'nurture' has polarized discussion to the point where the rational answers like '100 per cent of each' [22], or 'It is successional, therefore recursional and impossible to assign for the same reasons that development of any kind cannot be mapped onto DNA *or* external laws' [14] do not carry weight. People want very black-and-white answers to questions of heredity, and do not have patience with the expert who declares, as all experts do, that the matter is more complicated than that.

DNA preformationism provides a secure underpinning for just these reproductive worries. We can, if we believe in this simplified view, see the whole matter as a solid foundation for grounding of human foibles in a scientific, information-theory-justified framework. Just as the Christian belief in original sin gave the soul a colour—even if it was 'black'—so the belief that Man's ultimate nature is grounded in a complex, but understandable and listable, chemical sequence gives an emotional anchor within the perceived reproductive chaos. It makes motives and frailties not 'our fault', but determined by our fundamental DNA-blueprint nature. It allows us to forgive not only lability of others' emotions but also famine, war, pollution: as original sin was in our souls, so social and individual evil can be blamed on our ineluctable genes. As an icon, DNA-as-program serves a useful cultural, almost religious function. But it is not good biology.

It is comforting to imagine all those animals out there in the wild, living out their satisfying lives as they act out Aesop according to Grimm.

We enjoy wildlife films which show us this, and edit out most of the truth [23]. Real biological food chains are much less fun, much more difficult to feel happy about [24]. It is nice and 'hygienic' to think of the vase-shape of the iconic human female tract, easy to teach and with nice clean, easy-to-wipe-fingerprints-off, shiny models. The real thing is very threatening in all kinds of ways, with its associations with smells, excretion, all manner of 'dirty' associations ... pedagogy is very remote from sexuality. And DNA blueprints are a very satisfactory way of thinking about the complexity of human and animal bodies. The complexity is simply there, preformed, in the self-replicating DNA of egg *and* sperm, requiring only to be read out as a description of the organism. Real embryology, in contrast, requires lots of chemistry, physics, mathematics of dynamical systems [14], producing between them a real increase of complexity—so that the patch of cells on the surface of the yolk, from which the whole chick is made, really *is* less complicated than one growing feather. This biological way of looking at our heredity and development is like rationalism or humanism, compared with the blueprint model which is like the doctrine of original sin. It requires that we put our brains into gear, gives one a headache and an unwelcome sense of responsibility for the development of one's children—and of oneself.

Acknowledgements

Many of the ideas in this article have arisen, or been developed, in conversations with Ian Stewart. Mal Leicester made very useful criticisms of several drafts, refining my arguments and making the text more readable; I am very grateful for her patience.

References

1. Cohen, J. (1977). *Reproduction*. Butterworths, London.
2. Cohen, J. (1989). *The Privileged Ape*. Parthenon, Carnforth.
3. Shepard, P. (1978). *Thinking Animals; animals and the development of human intelligence*. Viking, New York.
4. White, T. H. (1954). *The Book of Beasts*. Jonathan Cape, London.
5. Opie, P. and Opie, I. (1974). *The Classic Fairy Tales*. Oxford University Press, London.
6. Luthi, M. (1982). *The European Folktale: form and nature*. ISHI, Philadelphia, PA.
7. Lodge, D. (1980). *How Far Can You Go?* Penguin, Harmondsworth.
8. Leeton, J., Trounson, A., Jessup, D., and Wood, C. (1982). The technique for human embryo transfer. *Fertil. Steril.*, **38**, 156–61.

9. Smith, R. L. (1984). *Sperm Competition and the Evolution of Animal Mating Systems.* Academic Press, London.
10. Baker, R. R. and Bellis, M. A. (1995). *Human Sperm Competition; copulation, masturbation and infidelity.* Chapman and Hall, London.
11. Cohen, J. (1990). The function of human semen coagulation and liquefaction *in vivo.* In *Advances in In Vitro Fertilization and Assisted Reproduction Technologies* (ed. S. Mashiach, Z. Ben-Rafael, N. Laufer, and J. G. Schenker), pp. 443–52. Plenum, New York.
12. Cohen, J. (1992). The case for and against sperm selection. In *Comparative Spermatology—Twenty Years After* (ed. B. Baccetti), Ares-Serono Symposia **75**, pp. 759–64. Raven Press, New York.
13. Cohen, J. and Stewart, I. N. (1991). The information in your hand. *The Mathematical Intelligencer,* **13**, 12–15.
14. Cohen, J. and Stewart, I. N. (1994). *The Collapse of Chaos; simple laws in a complex world.* Viking, New York.
15. Rucker, R. *Mind Tools.*
16. Carter, M. (1992). *Genetics and Evolution.* Hodder and Stoughton, London (or any other popular biology textbook . . .)
17. Cohen, J. (1967). *Living Embryos.* Pergamon, Oxford, 2nd edn.
18. Winfree, A. T. (1974). Rotating chemical reactions. *Scientific American,* **230** (6), 82–95.
19. Rothenberg, J. (1995). Ensuring the longevity of digital documents. *Scientific American,* **272** (1), 42–7.
20. Wolpert, L. *The Triumph of the Embryo.*
21. Toffler, A. (1980). *The Third Wave,* Pan Books, London.
22. Oyama, S. (1984). *The Ontogeny of Information.* Cambridge University Press.
23. *The Living Desert* (Disney Productions 1972) was typical; so was *Life on Earth* (BBC Enterprises Ltd, 1979)
24. *Trials of Life* (BBC Enterprises Ltd, 1990) was, initially, an attempt to show real biology, based on (1) above (Keenan Smart, personal communication, 1985), but was edited into nearly the Disney pretence.

JACK COHEN

Born 1933, could have become rabbinical but, having made some money by breeding tropical fishes, did Zoology at University College, Hull where he obtained a Ph.D. Did research in feather development, later transferring to hairs. Worked for the MRC and at Harvard Medical School. Lectured in embryology at Birmingham University. His book *Living Embryos* became a standard school text. A new idea about why there should be so many sperms was published in the '60s and carried an exciting research programme into the '80s. Worked with WHO on Immunological Control of Reproduction. Has designed science-fiction aliens for well-known authors. His most recent book (with Ian Stewart) is *The Collapse of Chaos.* He is Vice-President of the Linnean Society and a former chairman of Mensa. He is now at Warwick University.

Sticking up for adhesives

A. J. KINLOCH

Introduction

The problem with giving a discourse on the subject of 'adhesion and adhesives' is that everyone is familiar with the subject, since everyone uses 'glues'. Now some people are 'true believers' and will try using adhesives to join anything to anything, as is shown in a video sequence of television advertisements for 'Solvite' wallpaper-paste adhesive. These advertisements show 'Solvite' adhesives being used to stick wallpaper to a steam train-engine, to stick a man in a flying-suit to a wooden board and then suspending him over the City of Miami under a helicopter, etc. Obviously, some people have complete faith in the power of adhesives!

On the other hand, many people have no faith in adhesives at all, and such 'heretics' therefore take a 'belt and braces' approach when using adhesives. For example, Lady Macbeth clearly believed in always using additional mechanical fastening methods to ensure success, since she stated:

> We fail!
> But screw your courage to the sticking-place
> And we'll not fail.
>
> (Lady Macbeth to Lord Macbeth discussing the murder of Duncan; Act 1, Scene 7.)

However, I do hope that this discourse will convince you by 'Sticking up for adhesives' that we have much to gain.

The transport industry provides some excellent examples of the use of adhesives in critical and demanding engineering applications. Indeed, the use of adhesives in aerospace goes back a very long way, but not always with complete success. One of the earliest flying devices to use adhesives was the 'flying-wing structure' designed and built by Daedalus, possibly the engineer of the family. This structure was flown, of course, by his son, Icarus. He was possibly the chemist in the family, who selected the beeswax adhesive but clearly neglected to study the heat stability of this particular type of glue—a tragic oversight that cost him his life.

This introduces another theme in the story of 'adhesion and adhesives'. Namely, that it is a truly multidisciplined subject and requires the skills of the engineer and chemist, as well as the physicist, in order to achieve innovative and successful results.

Fortunately, more recent flying machines have used adhesives with far greater success. The early airships and biplanes of this century used adhesives based on casein, which is a natural material and is a by-product of milk. These adhesives worked well, except when they got wet. They then became very weak and smelt of old camembert cheese. However, engineers are a cunning breed, since it is claimed that they used this fact as an early form of non-destructive test. The aircraft engineers routinely smelt the bonded parts of the aircraft, and when the joints smelt of old camembert cheese they knew that the adhesive joints were about to fall apart, and that the adhesive should be replaced.

The problem of the poor aging of adhesives based on natural materials was largely overcome by the introduction of synthetic, polymeric adhesives. For example, in the Second World War the very successful 'Mosquito' aircraft relied mainly on urea-formaldehyde resin adhesives to bond together its wooden structure. This type of adhesive was, however, rather brittle in nature, and was prone to cracking and fracturing. Other types of early synthetic adhesive more suitable for bonding metals together, such as the phenolic resins, were also rather brittle materials. Thus, a major development in the 1940s was the modification of the chemistry of such very brittle, thermosetting resins to give tougher adhesives, and a very important type was based on a combination of vinyl-formal/phenolic-resin polymers. This invention represented a major development in adhesives technology and enabled metallic, as well as wooden, components to be bonded very successfully. For example, vinyl-formal/phenolic-resin adhesives were extensively employed by the designers of the 'Comet' jet airliners of the 1950s, particularly to give both high stiffness and strength, coupled with a relatively low weight, to the all-metal fuselage and wings.

These pioneering developments in the 1940s and 1950s by British scientists and engineers led to the construction of modern aircraft being dependent on the use of adhesives. The engineering adhesives used

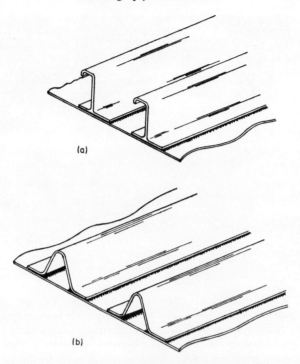

Fig. 1 Schematic diagram showing stiffening stringers which are adhesively bonded to skin panels: (a) extruded J-section stringers; (b) rolled-strip closed-channel stiffeners.

today are all based on synthetic polymers, such as modified-phenolic, epoxy, acrylic and urethane polymers. They are employed, for example, to attach stiffening stringers to the fuselage- and wing-skins, as shown in Figs 1 and 2, and to enable honeycomb structures to be made, as shown in Fig. 3. In these applications, the use of adhesives allows stiff and strong, but lightweight, components to be manufactured. It should be noted that in Fig. 2 the two gentlemen are undertaking the modern version of the 'old camembert cheese' test. However, their modern non-destructive test method uses ultrasonics, rather than noses, to detect whether the adhesive is satisfactory or not.

Helicopters also rely on adhesives and Fig. 4 shows a section of a helicopter blade. This blade uses several sections of stainless steel for the leading edge and a honeycomb trailing-edge (based on a plastic impregnated core and glass-fibre reinforced plastic skins), and all these different materials are joined together using adhesives. Although, whether the pilots are actually told that they are held up by 'glue' is doubtful!

Switching modes of transport, the 'Peugeot' rally car has a carbon-fibre reinforced-plastic drive shaft with the metal end-fittings bonded

Fig. 2 In the 'British Aerospace 146' one of the largest components is the wing-skin assembly which is manufactured using aluminium alloy. The stiffening stringers are bonded onto the skin using a modified-phenolic adhesive. The two gentlemen are conducting non-destructive tests on the bonded joints.

onto the drive shaft, as illustrated in Fig. 5. The adhesive, a toughened-epoxy polymer, has therefore to withstand the very high torque from the powerful engine of this rally car. Even faster are the latest designs of sports cars which are made from aluminium alloy to keep the weight of the vehicle relatively low. Such alloys cannot be readily spot-welded, and therefore adhesive bonding is employed as the joining method. The use of lightweight materials, such as aluminium alloys, plastics and fibre-composites, also leads to excellent fuel economy. Hence, the adhesive-bonding of these types of materials to manufacture cars, lorries, buses, trains, etc., is yet another rapidly developing area for the adhesives engineer.

Going on to two wheels, the US Olympic team recently used a racing bicycle constructed from lightweight aluminium–magnesium alloys with the various components being adhesively bonded, again with a toughened-epoxy adhesive being employed. Adhesives were also used in the manufacture of the bike ridden to victory by Chris Boardman in the last Olympic Games. His bike was made from carbon-fibre reinforced-plastic (CFRP) components which were glued together. Whilst on the subject of bicycles, a post-graduate student from a joint engineering-design course organized by Imperial College and the Royal College of Art recently

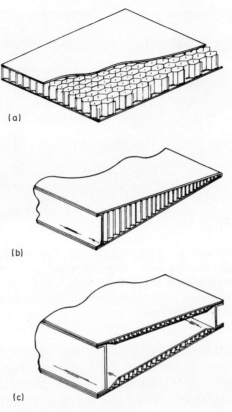

Fig. 3 Schematic diagram showing honeycomb structures which are manufactured by adhesively bonding skins to a honeycomb core: (a) honeycomb panel (flooring, etc.); (b) honeycomb structures for control surfaces, trailing edges, etc. (c) honeycomb structures for aerofoils.

designed and made a lightweight folding bike from aluminium alloy and plastic components. Needless to say, the various components were joined together using adhesives. Furthermore, at least one willing 'volunteer' from the audience had actually sufficient faith in adhesives to ride this bike around the lecture theatre. (And a successful bicycle ride by Dr B. R. K. Blackman, of Imperial College, was observed to take place.)

Now in all these many different applications of adhesives in engineering structures it is possible to identify three distinct stages in the formation of an adhesive joint. First, the adhesive initially has to be in a 'liquid' form so that it can readily spread over and make intimate molecular contact with the substrates, the substrates being the materials we wish to join. Secondly, in order for the joint to bear the loads which will

Fig. 4 A section of the blade for the Westlands 'Lynx' helicopter. The various materials used are stainless steel, plastic-impregnated paper honeycomb core and glass-fibre reinforced-plastic skins. These various materials are adhesively bonded to form the blade.

be applied to it during its service life, the 'liquid' adhesive must now harden. In the case of adhesives used in engineering applications, the adhesive is initially in the form of a 'liquid' monomer which polymerizes to give a high molecular-weight polymer. Thirdly, as engineers we must appreciate that the load-carrying ability of the joint, and how long it will actually last, are affected by (a) the design of the joint, (b) the manner in which we apply loads to it, and (c) the environment which the joint encounters during its service life.

To understand the science involved, and to succeed in developing the technology, we therefore require the skills and knowledge from many different disciplines. Indeed, we need the input from surface chemists, polymer chemists and physicists, and from design and materials engineers. Thus, the science and technology of adhesion and adhesives is a truly multidisciplined subject.

We have attempted to bring these different disciplines together by developing a 'fracture mechanics' approach to the failure of adhesive joints. The concepts of 'fracture mechanics' were introduced by A. A. Griffith in the 1920s whilst working at the Royal Aircraft Establishment, Farnborough. He recognized the importance of flaws in a material or structure. These flaws may be molecular-sized inhomogeneities, air bubbles, particles of dirt or dust, or they may be actual cracks. However they arise, Griffith proposed that the strength of a material, or structure, is governed by their presence. He proceeded to define a term, the fracture

Fig. 5 The upper shaft is a carbon-fibre reinforced-plastic (CFRP) drive-shaft, with adhesively bonded metal end-fittings, for the Peugeot rally car. The lower tube is manufactured from 'Kevlar'-fibre reinforced-plastic, again with an adhesively bonded metal end-fitting. The lower tube is a static tube in which the drive shaft rotates, so as to protect the CFRP shaft from impact damage.

energy, which is the energy needed to propagate a flaw through unit area of the material, or structure. The fracture energy is given the symbol G_c— where G is for Griffith and the subscript c indicates that it is the critical value for crack growth. The importance of flaws may be shown by loading, in polarized light, a material whose refractive index changes with load. If the strip of photoelastic material contains an edge crack, then the concentration of strain and stress around the crack is clearly seen by the intense pattern of colours which develop around the crack tip. I demonstrate how the ideas of fracture mechanics build on this fact by suspending a large sheet of paper from a horizontal support, with a dead load of about 300 N applied to the bottom edge of the paper sheet. I then use a sharp knife to make a small cut in one edge of the sheet of paper, and no adverse effect is seen. I then continue to cut deeper into the sheet, so making the edge-crack progressively longer. Nothing is observed to occur, until, at a critical length, the crack propagates extremely rapidly across the width of the loaded paper-sheet. The ideas proposed by Griffith allow us to calculate the size of the crack at which the rapid, catastrophic failure of the paper sheet occurs, and also to deduce the value of the fracture energy G_c needed for failure.

Now, we have been developing methods of fracture mechanics with respect to the failure of adhesive joints, so that we can determine the value of G_c for either a cohesive failure of the adhesive, or for an interfacial failure along the adhesive–substrate interface. The advances we have made in the fracture mechanics of adhesive joints have enabled us to understand better the science and technology of adhesion and adhesives, as I will attempt to show in this paper.

Interfacial contact and intrinsic adhesion

Introduction

As I mentioned previously, the first stage in the formation of an adhesive joint is concerned with attaining intimate interfacial contact between the adhesive and substrates, and then establishing strong and stable intrinsic adhesion forces across the adhesive–substrate interfaces. In order to achieve these requirements the substrates often have to be subjected to some form of surface treatment before the adhesive is applied.

Now, all these aspects emphasize the importance of surfaces and surface chemistry in the use of adhesives. However, surface science is a fiendishly difficult area of research, and the problems of understanding surfaces were summarized by W. Pauli by the following comment:

> God created solids,
> But surfaces are the work of the Devil!

However, we have persevered in our researches on surfaces, and we have used various experimental methods to understand the Devil's work, for example, the following:

(a) contact angles (i.e. the tangent angle at the contact point of a liquid droplet resting on a solid substrate surface) to determine the surface tension (or surface free energy) of the adhesives and substrates;

(b) X-ray photoelectron spectroscopy to identify the chemical nature of the surface;

(c) ellipsometry to measure the thickness of thin adsorbed adhesive or primer layers;

(d) reflection high-energy electron diffraction to gain information concerning the orientation of such layers;

(e) optical and electron microscopy to determine the surface morphology and topography;

(f) secondary-ion mass spectroscopy to detect the type of interfacial bonding which is present.

One problem for a university researcher is that to use most of these techniques requires a considerable expenditure, both in terms of capital equipment and running costs. It would be far less expensive for Imperial College if we changed our research areas from subjects such as adhesion science, physics, etc. to philosophy—at least according to Isaac Asimov, who once quoted an American University President as saying,

> Why is it that you physicists always require so much expensive equipment? Now the Department of Mathematics requires nothing but money for paper, pencils and erasers ... and the Department of Philosophy is better still. It doesn't even ask for erasers.

Mechanisms of intrinsic adhesion

However, expensive techniques such as X-ray photoelectron spectroscopy and secondary-ion mass spectroscopy have enabled us to determine exactly why materials do adhere. Many people, for many years, did believe that adhesion between the adhesive and substrates was due to some form of 'microscopic' mechanical interlocking. This theory essentially proposes that mechanical keying, or interlocking, of the adhesive into irregularities of the substrate surface is the major source of intrinsic adhesion. One example where mechanical interlocking is of prime importance is in the use of mercury amalgam for filling tooth cavities. The dentist drills out the tooth material to give a relatively large 'ink-bottle' pit, ideally with an undercut angle of about 5°, and a mercury-amalgam filling-material is then forced into this cavity. The main mechanism of adhesion which then occurs at the filling–tooth interface is mechanical interlocking. However, the attainment of good adhesion between smooth surfaces, such as adhesives to glass or to mica, exposes the mechanical interlocking theory as not being of general and wide applicability.

Since the intrinsic adhesion between the adhesive and substrates does not typically arise from mechanical interlocking occurring across the interfaces, then how does it arise? The answer to this question is that the adhesion arises from the fact that all materials have forces of attraction acting between their atoms and molecules, and a direct measure of these interatomic and intermolecular forces is 'surface tension'. The tension in surface layers is the result of the attraction of the bulk material for the surface layer, and this attraction tends to reduce the number of molecules in the surface region resulting in an increase in intermolecular distance. This increase requires work to be done, and returns work to the system on a return to a normal configuration. This explains why 'surface tension' exists and why there is a 'surface free energy'.

One well-known effect of surface tension acting in water is that it enables insects to walk on its surface. Another effect, which may be readily demonstrated, is that it enables a steel needle to be supported on the surface of water. However, when the surface tension is lowered, by the addition of a small amount of detergent, which decreases the forces of intermolecular attraction, the weight of the needle can no longer be supported by the surface tension, and the needle sinks.

Finally, it must be appreciated that solids, as well as liquids, possess a 'surface tension'; in the case of solids this property is generally termed the 'surface free energy'. Of course, the effects of a surface tension being present are far less readily observed in solids than in liquids.

Now, the forces of interatomic and intermolecular attraction may not only act in the bulk and surface layers of liquids and solids, but may also act across the interfaces between phases. Indeed, it is the presence of such forces of attraction which is generally responsible for the intrinsic adhesion between the adhesive and the substrates, and this most basic mechanism of adhesion was recognized by Michael Faraday over one hundred years ago! Thus, we can state that, provided sufficiently intimate molecular contact is achieved at the interface, materials will adhere because of the interatomic and intermolecular forces which are established between the atoms and molecules in the surfaces of the adhesive and substrates. The most common of such forces are van der Waals forces and these are referred to as 'secondary bonds'. Also in this category may be included hydrogen bonds. In addition, chemical bonds may sometimes be formed across the interface. This is termed chemisorption and involves ionic, covalent or metallic interfacial bonds being established: these types of bonds are referred to as 'primary bonds'. The terms primary and secondary are in a sense a measure, albeit somewhat arbitrary, of the relative strengths of the interatomic and intermolecular bonds.

To illustrate some of the above aspects I will use two examples taken from our current research work at Imperial. The first is concerned with the bonding of fibre-composite materials and the second is the development of organometallic primer layers.

The bonding of fibre-composite materials

As was shown earlier, the use of fibre-composite materials which are based on continuous glass- or carbon-fibres embedded in a polymeric matrix is steadily increasing in many engineering applications. Further, a recent development has been the use of a *thermoplastic* polymeric matrix. For example, matrices such as poly(ether-ether ketone) or poly(aromatic amides) have been developed and employed, as opposed to the more common thermosetting polymeric matrices based on an epoxy or an unsaturated-polyester resin. The advantages that the newer thermoplastic matrices can offer include shorter production times, higher toughness and easier recycling.

However, we discovered that the fibre-composites based on the thermoplastic matrices were difficult to join using conventional epoxy or acrylic engineering adhesives. Nevertheless, we found that subjecting the composite materials to a 'corona-discharge' treatment prior to adhesive bonding was a very effective method of obtaining good adhesion, and high joint strengths. The corona-discharge treatment of fibre-composite specimens is shown in Fig. 6. This treatment basically involves applying a high voltage (15–20 kV at a frequency of 15 to 20 kHz) across the air-gap between the composite surface and an electrode. The voltage is increased until it exceeds the threshold value for electrical breakdown of the air-gap, when the air is ionized. Hence a plasma, at atmospheric pressure, of excited oxygen, ozone, etc. ions and radicals is generated.

Fig. 6 Corona-discharge treatment of carbon-fibre reinforced-plastic (CFRP) specimens, where the CFRP material is based on a thermoplastic matrix. This pretreatment allows the fibre-composite materials to be successfully bonded.

These very active ions and radicals then react with the surface layers of the fibre-composite material, and so chemically modify its surface. The modified surface possesses a higher surface free energy; and this leads to better spreading of the epoxy adhesive over the surface of the fibre-composite and to higher intrinsic adhesion forces being established across the adhesive–fibre-composite interfaces. These aspects are reflected in tougher and stronger adhesive joints being made when the fibre-composite is corona-discharge treated prior to bonding.

To understand and quantify the effects of the corona-discharge treatment we have used many of the experimental techniques which I listed previously. For example, the surface analytical method of X-ray photoelectron spectroscopy may be used to detect the changes in the surface regions of the fibre-composite before, and after, surface treatment. The technique of X-ray photoelectron spectroscopy is based on placing the specimen in a ultra-high vacuum chamber and firing X-ray at the surface, but analysing the energies of the photoelectrons which are emitted. Since the photoelectrons can only escape from about the first 3 nm (1 nm = 10^{-9} m), then this method gives information about only the outermost surface regions of the material. X-ray photoelectron spectra of a carbon-fibre poly(aromatic amide) composite, before and after corona-discharge treatment, are shown in Figs 7(a) and 7(b) respectively. The increases in the concentration, and the type, of oxygen-containing chemical groups in the surface regions after subjecting the substrate to a corona-discharge treatment may be clearly seen. Further, contact-angle measurements showed that the epoxy adhesive would not spread over the untreated fibre-composites, but after subjecting the fibre-composite to a corona-discharge treatment the adhesive did spread readily over the composite's surface, and exhibited a very low contact angle. Thus, the increased presence of oxygen-containing chemical groups due to the corona-discharge

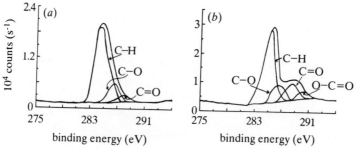

Fig. 7 X-ray photoelectron spectra of the carbon 1s peak of a unidirectional carbon-fibre/poly(aromatic amide) composite: (a) untreated; (b) corona-discharge treated (5 J mm^{-2} of energy applied).

treatment, as shown by using X-ray photoelectron spectroscopy, has increased the polarity of the surface of the composite and led to an increase in the surface free energy for the treated fibre-composite; and hence we obtain better spreading and intrinsic adhesion of the epoxy adhesive.

Now, it is difficult to demonstrate readily these aspects using the fibre-composites, which are black in colour. However, they can be illustrated using poly(tetrafluoroethylene) (PTFE) as a substrate material, which is white in colour. PTFE is better known by its trade names of Teflon and Fluon, and is of course the material used for the 'non-stick' coating of frying-pans and saucepans. The reason that PTFE is 'non-stick', and the reason why the thermoplastic composites are also difficult to bond, is that all these materials possess very weak surface forces and, therefore, possess very low surface free energies. I can demonstrate this by trying to spread blue ink over the surface of the white PTFE material. The ink does not spread over the surface of the PTFE to give a continuous film of ink, but remains as discrete droplets, which have a high contact angle. Even if I try to make the ink spread, by forcing it over the surface using a spatula, the ink still remains as discrete droplets. Now I can treat the PTFE by placing it in a solution of sodium naphthalenide for about 15 seconds. This treatment produces a dull brown-black surface layer, which has a far greater surface free energy than the untreated PTFE. These effects arise from the fact that the sodium naphthalenide treatment defluorinates the surface layer of the PTFE and introduces relatively polar, oxygen-containing, groups into the surface regions. (Also, it creates carbon–carbon conjugated double bonds in the surface regions, and it is these groups which cause the brown-black coloration.) The ink-wetting experiment may be repeated on the treated PTFE material, and the ink now readily spreads over the surface of the treated PTFE to give a smooth, continuous film of ink, which exhibits a very low contact angle. The greater adhesion of the treated PTFE may be shown by gluing together two PTFE strips, to form an overlap joint, using a cyanoacrylate (i.e. 'Superglue') adhesive. The joints are made and then left for a few minutes for the adhesive to polymerize, and so harden. It is demonstrated that the joint made using the untreated PTFE strips is very weak, and can easily be pulled apart. On the other hand, the joint made using the treated PTFE strips is sufficiently strong to resist all attempts to break the joint.

To summarize, the reason why surface treatments such as plasma treatments (e.g. the corona-discharge method) or chemical-etch treatments (e.g. the sodium naphthalenide solution) are effective is that they introduce chemical groups which are relatively polar into the surface regions of the substrate material. Hence, stronger intermolecular forces act in the surface regions of the materials, and this leads to an increase in the surface free energy of the substrate. Thus, we can attain (a) better interfacial contact between the adhesive and substrate, and (b) higher intrinsic adhesion forces of molecular attraction acting across the interface.

The use of organometallic primers

A second example of the importance of surface chemistry in adhesive bonding is the use of organometallic primers. The most common type of such primers are the silane-based primers. The types available commercially have the general structure $X_3Si(CH_2)_nY$; where $n = 0$ to 3, X is a hydrolysable group on silicon and Y is an organofunctional group usually selected to be chemically reactive with a given adhesive. The generally accepted mechanism of intrinsic adhesion for such primers is that they enable the formation of strong, primary, interfacial bonds across the substrate–primer–adhesive interfaces. They therefore effectively enable the adhesive to be chemically reacted, and so strongly bonded, to the surface of the substrate. This gives rise to strong intrinsic adhesion, which is reflected in strong and durable (i.e. water-resistant) adhesive joints.

Now we have been developing such primers which form only a 'monolayer' in a joint. As the name suggests, a 'monolayer' is where only one molecular layer of a chemical species adsorbs onto a surface, and for many years monolayers have been known to be powerful agents for modifying surfaces. We can see what a great influence monolayers may have from an experiment where I ignite a dish containing a saturated aqueous solution of ether and then extinguish the flames by using oleic acid (i.e. $CH_3(CH_2)_7CH{=}CH(CH_2)_7COOH$). The oleic acid adsorbs as a monolayer via its highly polar acid-end group. Since oleic acid does not burn, it smothers and so extinguishes the flames. (The experiment, fortunately, worked as planned—thus the fire-brigade were *not* called upon!) As can be clearly observed, the amount of oleic acid needed to extinguish the flames is very, very small. This is because only one molecular layer is needed and so the thickness of the oleic-acid layer is about 3 nm; when considering the thickness of this monolayer bear in mind that 1 nm = 10^{-9} m, one millionth of a millimetre.

Now to achieve a monolayer of an organometallic primer adsorbed onto the surface of a substrate, simply from dipping the substrate into a solution of the primer, we have synthesized a long alkyl-chain based silane, namely 18-nonadecenyltrichlorosilane, which has the chemical formula

$$CH_2=CH(CH_2)_{17}SiCl_3$$

The substrate, an aluminium alloy, is immersed in a 0.1 M solution of this silane primer in 90 per cent hexadecane+10 per cent chloroform. The silane first hydrolyses, to give 18-nonadecenyltrihydroxysilane, and then chemically adsorbs onto the surface of the aluminium alloy, with the hydroxyl groups reacting with similar groups which are present on the aluminium oxide to give a —Si—O—Al— chemical bonds. The presence of the long alkyl chain forces the hydrocarbon chains to pack tightly together and, being non-polar, they orientate themselves away from the substrate. Thus, an orientated monolayer of the silane primer, about 3 nm thick, and which is chemically bonded to the aluminium oxide, is formed on the surface of the aluminium alloy. The second step is to convert the vinylic-end groups on the adsorbed primer to groups which may react with the adhesive which is to be used to bond the silane-primed aluminium-alloy substrates together. In the present experiments the vinylic-end groups are converted to hydroxyl groups. (It should be noted that, if the vinylic-end groups were converted to hydroxyl groups *before* the primer was adsorbed onto aluminium alloy, the presence of polar groups at both ends of the primer molecule would have prevented the formation of an orientated monolayer of the silane primer.) A polyurethane adhesive is then used to join the silane-primed aluminium-alloy substrates.

The measured fracture stress of the joint, as a function of the length of the sharp crack which is placed at the interface in the joint, is then determined; the results are shown in Fig. 8. As may be seen, when the fracture stress is plotted as a function of the inverse of the square-root of the crack length, an excellent linear fit is found for any given type of joint, as demanded by the theory of fracture mechanics developed by Griffith. For the joints where the vinylic-end group on the primer was not changed, so giving a primer which would not react with the incoming adhesive, the adhesive fracture energy G_c is lower than that of the control joints; where no primer was employed. However, when the vinylic-end group on the primer was changed, so as to give a primer which would now react with the incoming adhesive, the joints possessed an adhesive fracture energy G_c which was significantly greater than that of the control, unprimed joints. Thus, the presence of an orientated monolayer of the organometallic silane primer (which is therefore an inherently strong interlayer) and which can chemically react with both the aluminium oxide and the adhesive (to form a primary, chemically bonded inter-

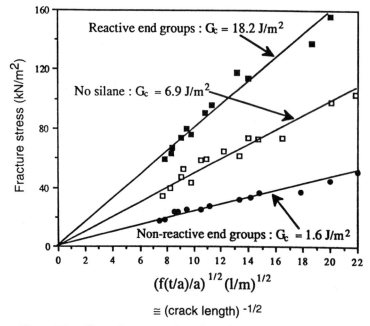

Fig. 8 The effect of using a silane-based primer in polyurethane adhesive/aluminium-alloy adhesive joints. The figure shows a plot of the fracture stress versus a function of the inverse of the square root of the length of the crack inserted at the interface. The value of the adhesive fracture energy G_c may be obtained from the slope of the linear relationship for a given joint. (The control joints employed no primer.)

layer) dramatically increases the toughness of the joint. Further, if the joint is immersed in water, or exposed to a high relative humidity, then the presence of such a primer may also greatly increase the resistance of the interface of the joint to attack by moisture. Thus, the durability and service life of the joint may be significantly increased.

Hardening the adhesive

Introduction

The second stage in adhesive bonding involves 'hardening' (sometimes called 'curing') the adhesive. This is necessary since, to achieve spreading of the adhesive over the substrate and establish interfacial molecular contact and adhesion, the adhesive has to be in a 'liquid' form. However, for the adhesive joint to be able to bear loads, the adhesive must now harden to form a relatively strong, rigid solid.

This brings us to the role of polymer chemistry and physics in the science and technology of adhesion and adhesives, and ***we*** physicists can now put the rest of the science and engineering community in their rightful place. At least according to Rutherford, who once stated:

> The only true branch of science is physics, the rest is just like collecting postage stamps.

On the other hand, ***we*** chemists did fight back against this attitude, thanks to Frederick Soddy who said

> Chemistry has been termed by the physicist as the messy part of physics, but there is no reason why the physicists should be permitted to make a mess of chemistry when they invade it.

Chemical aspects

The importance of hardening the adhesive can be readily demonstrated by trying to bond together two pieces of balsa wood, using water as the glue. Although the water spreads readily over the balsa wood, the strength of the overlap joint between the pieces of wood is obviously low, since the liquid water has no significant *bulk* strength. If we repeat the experiment, but now freeze the water by dipping the joint in liquid nitrogen for a few seconds, then whilst the water is frozen—so the adhesive is now ice, of course—the joint is relatively strong. Indeed, when I break the joint it actually fails in the balsa wood (i.e. substrate) away from the bonded region and not in the adhesively bonded overlapped region.

Now in the case of engineering adhesives we need a hardening method which is not as temperature sensitive, and which gives more durable joints, than 'freezing'. And most engineering adhesives harden via the formation of a crosslinked, thermosetting polymer. An excellent example of this polymer chemistry in action, and one which many of the audience will have used themselves, is 'Two-Tube Araldite'—a product sold by all good 'Do-it-Yourself' shops! You simply first mix together the resin (which is an epoxy resin) from one tube and the crosslinking agent (quite logically called the 'hardener') from the second tube. The adhesive is then used whilst it is in the 'liquid' form, so that you can spread it over the substrates. At this stage the joint has virtually no strength, since the adhesive is still a 'liquid'. However, the adhesive then hardens over the next few hours, which is achieved by the epoxy resin reacting with the hardener to form a crosslinked, thermosetting polymer. Once hardened, quite high loads can be applied to the polymeric adhesive, without the adhesive undergoing plastic deformation (i.e. without the adhesive flowing) or fracturing.

Multiphase adhesives

For adhesives to be used in very demanding engineering applications, the chemistry of the hardening process is often designed so as to give an adhesive which possesses a 'multiphase' microstructure. Indeed, the formation of a multiphase microstructure is crucial in enabling adhesives to be used as very tough, engineering materials. Typically, a liquid rubber is dissolved in the 'liquid' (i.e. uncured, monomeric adhesive) and as the adhesive polymerizes (and so hardens) the rubbery component phase separates and forms small spherical particles of rubber in the adhesive material. The small rubber particles can be clearly seen in a transmission electron micrograph, Fig. 9, and they are about 1 to 5 μm in diameter ($1 \ \mu m = 10^{-6} \ m$).

To demonstrate the effects of toughening the adhesive by the inclusion of such rubber particles I can bond together two strips of aluminium alloy. For one joint I use a simple, i.e. non-toughened, adhesive whilst for the other I use the same basic adhesive but which now possesses an internal microstructure of rubber particles, as shown in Fig. 9. In the case of the non-toughened adhesive, the joint breaks easily when the lap joint is bent. However, for the toughened adhesive, the lap joint

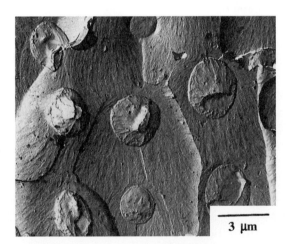

3 μm

Fig. 9 A transmission electron micrograph of a rubber-toughened epoxy adhesive. The spherical rubber particles in the matrix of cured epoxy polymer may be clearly seen.

can be bent repeatedly backwards and forwards, without the adhesive failing.

Now, these experimental results do clearly show that the rubber-modified, multiphase adhesives are indeed extremely tough. However, I have always tried to follow Sir Arthur Eddington's rule:

> It is a good rule not to put too much confidence in experimental results until they have been confirmed by theory.

Therefore, over the last few years we have developed theoretical models to explain why such adhesives are so very tough.

We have theoretically modelled (a) the microstructure, (b) the toughening mechanisms, and (c) the resulting fracture energy G_c using finite-element stress analysis. Plate 35 shows a computer-generated model of two rubber particles; we only need to model one quarter of each particle because of the symmetry of the structure. Between the rubber particles is the cured adhesive, in this case we have fed into the model the mechanical (i.e. the stress versus strain) properties of a typical crosslinked epoxy polymer. Next we can apply a load to our model, and the colours seen in Fig. 10 represent different levels of stress which result from the interactions of the microstructure and the applied load. The band of colours running at approximately 45° between the rubber particles shows the development of one of the many plastic shear bands which are initiated by the presence of the rubber particles and form in the epoxy adhesive. These plastic shear bands absorb mechanical energy and so reduce the stress concentrations at the tip of any crack, or flaw, in the joint. Hence, the presence of the rubber particles greatly increases the toughness of the adhesive. From the computer model we can theoretically predict the level of toughness of the multiphase adhesive for a given microstructure, providing we measure the stress versus strain behaviour of the rubber and epoxy phases so that we can feed these properties into our model. The values of the fracture energy G_c for a rubber-toughened epoxy adhesive from the theoretical modelling studies are compared with the experimentally measured values in Fig 10. Clearly, following Sir Arthur Eddington's rule, we may now have confidence in the experimental results. Such modelling methods are currently being used in order to develop even tougher adhesives in a more efficient, and less empirical, manner.

The development of these multiphase adhesives represents another major development in adhesives technology, and many industries now rely on such materials. For example, the 'Jaguar XJ220' sports car employs such adhesives extensively for joining the aluminium-alloy parts used in its construction. Indeed, if you could spare about £400 000, then you could help in 'Sticking up for adhesives' by buying one of these vehicles!

Predicting the strength and service life of adhesive joints

Introduction

Turning now to the third and last stage, the design of the joint, the way in which loads are applied to it and the service environment that it must withstand will all affect its mechanical performance and service life. So, now the skills and knowledge of the materials engineer are demanded.

For example, the combination of hot/wet climatic conditions and cyclic fatigue loads represents a very demanding environment for an adhesively bonded joint. Indeed, there are cases where engineering components have prematurely failed under such conditions after they have been in-service for periods of just a few months. Obviously **we** engineers wish to be able to predict, and prevent, the failure of adhesively bonded components. Therefore, accelerated aging tests are often undertaken during the design phase of an engineering project. Accelerated aging tests typically involve exposing bonded joints in water, or the environment of interest (e.g. a corrosive salt-spray), at a relatively high temperature, for example, maybe six months exposure in boiling water, or at least water at say 60°C. A major problem which may be encountered with such an approach, and a reason why such accelerated tests may be very misleading is succinctly summed up by the question

When did boiling an egg ever produce a chicken?

Thus, it is important to ensure that accelerated tests are selected which do give the same outcome (i.e. the same mechanisms of aging) as would be seen in real life—the aim being to accelerate the mechanisms of environmental attack, not to produce mechanisms different to those seen in real life.

The fatigue behaviour of adhesive joints

We have been working to predict the lifetime of bonded joints which are subjected to cyclic fatigue loads. Such loading occurs when the joint is subjected to oscillating loads, and this type of loading may be particularly damaging to all types of materials and components. A fracture mechanics test has been used to generate the data shown in Fig. 10. Here the rate of crack growth per cycle, da/dN, is plotted against the maximum adhesive fracture energy G_{max} which occurs in a loading–unloading cycle. It may be seen that the lower the value of the maximum load in

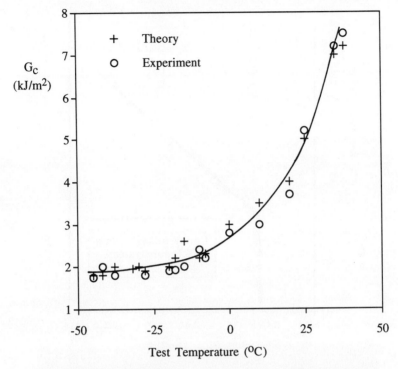

Fig. 10 Values of the fracture energy G_c of a rubber-toughened epoxy plotted as a function of the test temperature. Note the good agreement between the theoretical predictions from the modelling studies and the experimental measurements.

the cycle, and hence the lower the value of G_{max}, then the slower is the rate of crack growth through the joint. Further, there is a threshold value of G_{max}, below which no crack growth occurs. This threshold value obviously provides a very useful design limit when used with appropriate safety factors. Providing the maximum load applied to the bonded component is always below the corresponding value of this threshold value of G_{max}, then no failure due to fatigue loading should ever be observed.

Now, these fracture mechanics data shown in Fig. 11 may also be employed to predict the actual fatigue life of other designs of adhesive joints and bonded components. The basic approach is to derive a relationship for the results shown in Fig. 11, and then integrate this relationship. For example, the data may be used to predict the number N_f of cycles needed to cause failure of a lap joint, which consists of two strips of fibre-composite material overlapped and bonded together, when subjected to a given cyclic fatigue load. Indeed, we have obtained good agreement between the experimental results and the theoretical predic-

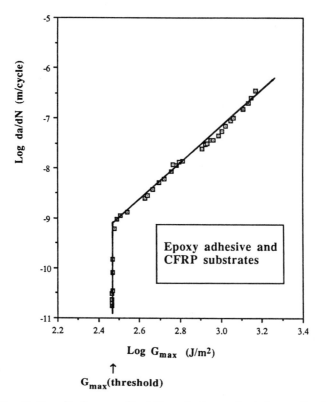

Fig. 11 Results from cyclic fatigue tests conducted on adhesive joints which consist of an epoxy adhesive bonding carbon-fibre reinforced-plastic substrates. The crack growth rate, da/dN, per cycle is shown as a function of the maximum fracture energy G_{max} in any loading–unloading cycle. Note the presence of a lower bound, threshold, value of G_{max}, below which no cyclic crack growth will occur. (Frequency of testing 5 Hz.)

tions for the number N_f of cycles such joints may be subjected to before failure is observed. We are currently developing such methods to enable us to predict accurately the lifetimes to be expected from different designs of bonded components.

Conclusions

I trust that the examples given in this paper have greatly increased your faith in the use of adhesives. But I now come to a final, and extremely convincing, reason for 'sticking up for adhesives'—which is that nature, invariably, has anticipated the efforts of mankind. Indeed, nature has most successfully acted as **both** the chemist and the engineer for many, many thousands of years.

An excellent example of nature using adhesion and adhesives is illustrated by the survival method of the plant *Drosera rotundiflora*, otherwise known as the sundew plant. This plant survives by digesting the body fluids of dead ants, but first, of course, it has to capture the ant. Now, to capture the ant it *does not* drive a nail through the ant, or use mechanical fasteners. It does, however, rely on 'glue'. At the end of each stalk on the plant there is a globule of glue. Photographs of the capture of the ant show that the glue spreads readily over the body of the ant, making a low contact angle, and the glue is very tough. As in the case of engineering adhesives, these factors combine to give excellent adhesion and joint strength. However, nature's efforts clearly score over those of mankind, since the sundew plant does not, of course, have to pretreat the body of the ant before its glue can adhere, not did it rely on mathematical formulae and modelling to develop its glue.

However, if you are still not sufficiently convinced to become a convert and have faith in adhesives, and so join me in 'sticking up for adhesives', then I would suggest that you follow the advice of J.E. Gordon:

> When all else fails
> Use bloody great nails.

Acknowledgements

The lecturer would like to acknowledge the assistance of Alcan, British Aerospace, Ciba Composites and Polymers, Ford Motor Co., Lotus, Permabond Adhesives, TWI, University of Surrey, Westland Helicopters, Williams Grand Prix Engineering, many colleagues at Imperial College and Mr W. A. Dukes, formerly of the Ministry of Defence, for their assistance with exhibits, demonstrations, etc. used for the discourse and accompanying exhibition.

Bibliography

Bogue, R. H. (1922). *The Chemistry and Technology of Gelatin and Glue.* McGraw-Hill, New York.

Bonded Structures Division (1957). *Bonded Aircraft Structures.* Bonded Structures Division, Duxford.

De Bruyne, N. A. and Houwink, R. (eds.) (1951). *Adhesion and Adhesives.* Elsevier, Amsterdam.

Eley, D. D. (ed.) (1961). *Adhesion.* Oxford University Press, London.

Kinloch, A. J. (1987). *Adhesion and Adhesives: Science and Technology*. Chapman and Hall, London.

Wake, W. C. (1976). *Adhesion and the Formulation of Adhesives*. Applied Science Publishers, London.

ANTHONY J. KINLOCH

Born 1946 in London, he received his first degree in Polymer Science and Engineering, and then in 1972 a Ph.D. from Queen Mary College, University of London, for his research on the mechanics and mechanisms of adhesion. He then started work for the Ministry of Defence, and undertook research and development in the area of adhesion, adhesives and polymer science. In 1984, he joined the Department of Mechanical Engineering, Imperial College as Reader in Engineering Adhesives, and in 1990 was appointed Professor of Adhesion. He has served on many SERC, Institute of Materials and Government Committees, and has been a visiting Professor at Universities in Europe and the USA. In 1988 he was awarded a D.Sc. (Eng.) from the University of London for his research in Materials Science, and in 1992 he received the Award for Excellence in Adhesion Research from the American Adhesion Society.

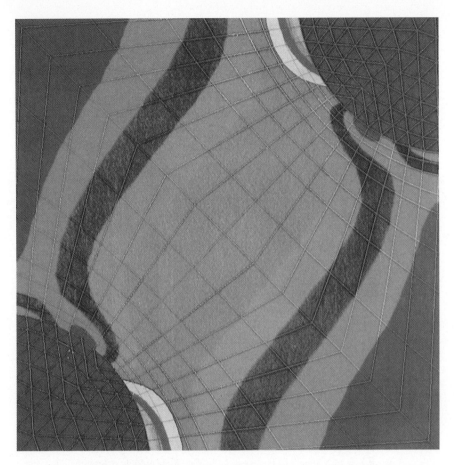

Plate 35 A computer-generated model of the deformation behaviour of a rubber-toughened epoxy adhesive. The finite-element analysis shows two (one-quarter) spherical rubber particles. The material between the rubber particles is the epoxy adhesive. The green-coloured band represents the development of a plastic shear band in the epoxy polymer—initiated by, and running between, the rubber particles. (The adhesive is being loaded vertically.)

Plate 36

Bang

Marden Lodge Sch
Croydon Road
Caterham Surry
CR3 6QE
5ᵗʰ May 1994

Dear
Doctor Hubbard,
I'm writing to
thank you for coming to Marden Lodge
School. I really enjoyed your experiments,
especially the illuminous yellow and
green in the test tubes. It was very
interesting and the way the glove got
bigger and bigger. When I got my dad got
out my chemistry set and we made a
crystal with it. I would like to study
chemistry at my high school. Thank
you again for coming.
From
Neil
Warren.

Plate 37

To Planet Earth with love from chemistry

ANN HUBBARD

For some years now I have been lecturing to young people about chemistry, at the same time using lots of illustrations and demonstrations, very much in the tradition of Michael Faraday and the Royal Institution. Here, I want to give some flavour of how I go about catching and holding the interest of young people in what can sometimes be quite complicated phenomena. The starting point is the world around us as we see it.

> Welcome to Planet Earth—a place of blue nitrogen skies, oceans of liquid water, cool forests and soft meadows—a world positively rippling with life [1]

'We are going to talk about our planet—Earth. If you look at it, it is really only made up of three things—the blue stuff (the children tell me it's the water or sea or ocean)—and the brown stuff? (it's the land) and what is the very thin layer that goes all the way round the outside? It starts at your feet and only goes up for a few miles? (it's the atmosphere)'. Here I want to convey how to use accessible language and questions to involve my young audience—if a sophisticated word like 'atmosphere' comes back from any eight year old they are rewarded with 'you are a budding scientist'. I know they have to feel involved in what I am communicating. Questions, even rhetorical ones, do this: 'Where did our planet come from? Why should we love and care for our planet Earth?' I will keep quoting directly what I say to the young people because I feel strongly that it is the simplicity of language as well as enthusiasm that makes the audience want to keep listening.

Whenever I give a lecture to young people, usually to non-scientists, I do not like to disappoint them. For most, an awareness of the molecular structure of everything is low: it always amazes me how low. Nevertheless, the expectation that a Chemistry lecture will include explosions is

Fig. 1

very high! 'Ere Miss, are you gonna do an explosions?' is often the request. I do not disappoint them, because Chemistry is fun, it is tremendous fun so as a taste of things to come, 'don't blink or you will miss something' and after a warning, a loud BANG starts the lecture as the balloon of hydrogen explodes. In my experience children enjoy the 'bangs' more than adults (Fig. 1). When the hubbub among the audience subsides I can make the poor excuse that the hydrogen explosion serves to introduce the idea that 'millions and millions of years ago a star exploded.—The atoms in the star travelled across the universe, eventually collecting around the star that is our Sun and became the planet Earth.' Later there were 'plants, animals and trees and eventually you— did you know that you were made of stardust from millions of miles across the Universe?' I like this idea, and I know the girls particularly like being told they are made of stardust. 'How does this stardust make you and everything else on the planet? To understand that you have to understand Chemistry, because Chemistry is about the tiny little molecules that make up everything.' These are not, in my experience, statements of the obvious.

'Here is a model of a tiny water molecule—that everyday stuff that sloshes about. If you stop it sloshing out and get it to stick together it builds itself into an ice cage and slowly into a snowflake.' The models I use of ice and diamond (Fig. 2) relate the molecular structure to their

Fig. 2

everyday appearance, a link often lost on people. Most people do not realize that it is knowledge of the molecular world that brings an understanding of the world we see, touch and taste.

'Lumps of rock are molecules sitting still. They don't sit completely still, the atoms in a solid can wobble about as the "wobbly" crystal model shows. Some of these solids can be very beautiful—look at a crystal or a gemstone. The movement of the liquid bit of the planet is shown as waves in a bottle.' This bottle, containing half blue coloured water and half petroleum ether, travels around the audience so they can appreciate the trembling molecular motion of the two immiscible liquids flowing past each other (Fig. 3). Plastic spheres are poured out as liquids and thrown among the children as moving gas molecules. These are greeted with delight and the concepts of molecular motion are illustrated. Yet scientists take this simple idea for granted. When, in normal experience, except for water, do we see a substance changing from gas to liquid to solid? The 'sitting still' of a liquid to make a solid is normally hidden in a freezer. We show a liquid poured from a beaker to build a large 'crystal castle' (supersaturated sodium acetate crystallizing), thus lining up the molecules and 'sitting them still' (Fig. 2).

'We can trap the gas molecules that are normally buzzing around you, so you can see them, simply by making them very very cold so they sit fairly still.' Liquid air (or more precisely liquid nitrogen) is visible in a

Fig. 3

glass Dewar flask and 'if you pour it out it buzzes off around your toes and noses. How cold is liquid nitrogen? What number would the weather men put up on a cold winter's night?' The childrens' answers come back '4', I say 'colder', '0', 'even colder', the under 12s have to stop and think! Then, I get 'minus', 'minus how many?' and eventually— '−10!' To get a feel for the temperature scale I go on and say 'this liquid nitrogen is not −10° or −50° it is −196° and this probably makes it the coldest thing you have ever been next to in your lives.' To most people it will be—it is even to me. 'It is so cold it does very funny things to other molecules, take bouncy rubber tubing; hit it with a hammer and it bounces back, put it in liquid nitrogen and all the bounce energy is lost, it shatters just like a piece of glass.' I do not think we realize that unless we are scientists (about 10 per cent of the population) we are unlikely to have seen this (Fig. 4). 'So scientists, who are not peculiar people, but have to go home and have tea and go to bed, use liquid nitrogen to store molecules so they cannot move and change.' A big bag of ammonium nitrate fertilizer and a loaf of bread also convey how nitrogen from the air is a source of this fertilizer so we are well fed 'but not everyone is so fortunate. If we can understand the molecules we can make life better for everyone.'

'We have other gas molecules in the air—carbon dioxide which when made very very cold and sitting still is dry ice.' Dry ice is of course

Fig. 4

an incorrect name because these are not water molecules and it is extraordinarily difficult to undo this misnomer. 'If we use hot water to warm up the carbon dioxide it has energy to move and spread out as a gas.' Now clouds literally billow across the table and down over the floor (Fig. 5). Then I carry the beaker round so they can 'touch' molecules of carbon dioxide! This causes great excitement. Now I can ask them 'what do you notice that is unusual' because the carbon dioxide molecules are going down and 'forming a gas tablecloth' they can always recognize the downward movement and then can understand that we can use a 'big blanket of carbon dioxide from a fire extinguisher to put the fire out'. Seeing a plastic glove fill up with carbon dioxide so it looks like a giant cow's udder is equally entertaining.

These simple experiments, well known and enjoyed by chemists the world over, still have not been seen by *most* children and their parents.

'There are molecules in the air that you need to breath—oxygen, how much of the air is oxygen? Not much more than 20 per cent would cause all sorts of trouble because oxygen joins up with other molecules in the burning process.' Now I show things 'joining up' with oxygen, emphasizing they are all from the planet, like phosphorus or sulphur from volcanoes. The eery blue flame of sulphur and the brilliant white flame of phosphorus burning in pure oxygen light up the darkness. 'The white fumes of sulphur dioxide make you cough. So when we burn fossil fuels

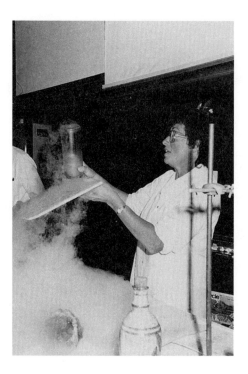

Fig. 5

in power stations we have learned not to let the sulphur dioxide gas out into air to cause acid rain.' By showing we understand what molecules do we can care for the planet, unlike the Victorians who did not have this knowledge. 'We can stop things like industrial acid rain and protect the plants.' I am consciously trying to counteract Chemistry's unpopular and 'difficult' image.

'You use oxygen to keep you alive and breathing.' People do not realize how much energy is released from food. 'Have you had two teaspoonsfull of sugar today?' If you watch the fast release of energy from sugar burning with chlorates (V), it demonstrates the large quantity of energy released and at the same time the reverse of photosynthesis! I have had many nutrition lecturers want to learn to use this demonstration. To keep the adults intrigued we pose the problem of the burning candle in a stoppered flask. When dropped, in the microseconds of the fall the candle goes out. (Of course we always hope to catch the flask.) We ask them why does the candle go out? It is by trying to explain things like this that scientists ask questions and seek to understand the wonder and the magic of their world. 'All sorts of unexpected things happen when you start looking at things and asking questions.'

Now we go on to the rocks and minerals and show a reaction that is

Fig. 6

'older than Jurassic Park!' Current relevance is all important. 'This happened before the dinosaurs walked the earth and it is the reason you cannot go round the planet and pick up natural lumps of metals like iron and aluminium. These are locked up in the rocks and minerals. When molten metal comes out of a volcano it is hit by oxygen in the air.' (This is poetic licence in the scientific sense.) The audience watches tiny pieces of iron powder sparkle through the flame and aluminium dust flare violently when blown into the flame; a reaction that would amaze a barbecue cook using aluminium foil (Fig. 6). The metals have joined up with the oxygen to make rocks and minerals.

'Think for a minute of all the things made with metals. Without knowing how to extract the wealth of metals there would be no copper wire to conduct electricity, no television, no videos, no computers, no cars, no planes.' A thermite reaction illustrates the extraction (iron oxide and aluminium powder) with molten iron pouring from the hole in the bottom of a flower pot at 1500 °C 'it is probably the hottest thing you have been next to in your life!'

'What about the sea—everyone knows one chemical from the sea! From salt we can make a green gas called chlorine. It is safely shut up in a flask because it is a very poisonous reactive gas.' In the lower flask under vacuum is a copper alloy (dutch metal)—'copper is like gold which sits around for ages and ages and is perfectly alright.' As the chlorine gas is sucked down onto the copper the chlorine joins up with copper, bursting into flame and leaving a green powder. 'Chlorine is used for you everyday. We live in a country where, when you turn on a mains water tap you can get water which is guaranteed, by Act of Parliament, to be completely safe to drink because it is bacteria free. Many people worldwide are not so fortunate. Also, think about the current fashion for supermarket bottled water—not necessarily bacteria free and at a local water board's estimate sometimes 3000 times more expensive per litre! The mains water is 'germ free' because it is treated with chlorine before it goes into the pipe.' Also, imagine an operation without an anaesthetic (halo-compounds). Your great grandparents had to make do with a leather strap and a bottle of whisky [2]. This relevance of Chemistry to everyday lives needs pointing out—people, old or young, may not remember the details but they go away with the idea that the subject is literally tangible.

Salt from the sea is a source of another gas—it used to be called 'spirits of salt' because a white mist rises eerily from a mixture of salt and concentrated acid. The youngsters, even little ones, have heard of acids, so hydrogen chloride or spirits of salt [3] to them is 'acid salt gas'. I use it in a classic fountain experiment and say the acid gas molecules 'love to be in between water molecules, as they dive in they leave a space which is filled up by more water molecules fountaining up.' It surprises me how many of today's undergraduates have not seen this demonstration of solubility and of a vacuum being filled by the molecules moving into the space. Sadly, the art of demonstrating seems to be dying. I consider many of my experiments as ordinary, everyday, but young teachers often do not have them as part of their repertoire. Chemistry is not only finding out about the planet but also how we can use the molecules to make things, even using things as simple as solutions of salts 'made from dissolving rocks and minerals in the rain'. This is of course an over simplification but it depends on the level of knowledge of the audience. Redox and complexing reactions would only be acceptable to an audience of upper sixth form 'A' level chemistry students, graduates, and qualified chemists.

Now we can also mix these solutions. The first, containing 'iron', is pale yellow; this is mixed with a clear colourless solution giving us 'new molecules—some red ones just like blood'. The second 'has got some copper in it and is mixed with molecules made from nitrogen from the air' giving an amazing dark blue colour. To the chemist these are the iron thiocyanate complex and the copper ammonia complex. To the children it is just magic that I can make red, blue, green, orange and yellow so easily, all from 'dissolved rocks' that are metal compounds. Brightly coloured pictures, works of art (cadmium yellow, cobalt blue) depend on the paint box substances made from rocks and minerals. It never ceases to amaze me how much children remember. I have taken a very short version of this lecture demonstration to a school for children with severe learning difficulties. I returned a year later to a class, half of whom had seen it the year before. Their experienced teachers said 'don't worry they won't remember a thing.' I will not make that mistake again—they got really excited bouncing up and down in their seats like yoyos as they knew what was coming next. If these children remember, imagine the impact on the budding scientist. To make new molecules means we have to understand how they join together—a 'tornado in a bottle' is a whirling storm where they can actually see the way the molecules are mixing (mercury(II) and iodide). Not only must we understand what the molecules are, but we also have to control them safely. 'How many of you have heard of a chocolate biscuit factory exploding?' 'How many chocolate factories are there?—Imagine yourself standing in a super-market, look around at all the things—all molecules—that have been made in factories working day and night. How often do you hear of a factory exploding? The molecules have to be kept under control and safe.'

Two reactions under control are classical iodine clocks. We mix two flasks simultaneously and ask the children to say when they see some-thing happen in each—'not before because scientists have to be accurate and observant!' The flasks go blue-black together (unless my assistant John has been cheating and warming his!) and this instantaneous change is greeted with surprise and delight, as anyone who has shown this to a class for the first time knows. The oscillating blue and red circles of a different reaction are another beautiful example of molecular control.

'Sometimes reactions go out of control and that means we must under-stand what the molecules are doing.' They never mind seeing the hydro-gen balloon go BANG again as an 'out of control reaction'. By controlling the explosion's direction the energy is used to launch a lemonade plastic bottle as a rocket (hydrogen–air mixture). This is probably our most popular demonstration. The children get excited as we produce a second or third bottle—after all, all good experiments should have reproducible results! Unlike us, today's kids take phone calls via satellite to Australia and satellite television for granted, and it is good to emphasize the posi-tive nature of explosive reactions.

A warning about trying hazardous experiments is driven home by doing a really out of control explosion. A methane–oxygen mixture is a reaction 'that sadly could happen in your home and is why the gas board rush round to sort out a gas leak'. People live in such a protected

environment I think there is a positive advantage in showing them hazards. The plastic bottle is small, the warning is large and the following
explosion is enormous, showing that 'we do have to manage the
molecules carefully.'

'If you have launched a rocket going to Mars there is no air in space to
provide oxygen to burn the fuel.' So we ignite alcohol as a fuel from a
solid source of oxygen (chromium(VI) oxide) without a match to show
energy release without the activation of a flame.

Hopefully the children begin to think, like us, that these molecules are
amazing and also realize that we are only making them do what they do
naturally—so this world is open to them—they could do this too, when
they understand the molecules.

Their world of cars, planes, television, clothes, sports equipment, and
medicines depends on chemical molecules including 'oil—a treasure
chest of molecules'. There is only time to do a few experiments from this
treasure chest. Mixing 'oil molecules' to turn a piece of white cloth into a
bright orange cloth shows that 'many of the bright colours' are made
with these diazonium dyes. Early Victorian clothes in drama productions illustrate the lack of colour that existed before these reactions were
discovered. The propensity of the young for black sometimes makes me
wonder whether colour will ever be back in fashion—but at least on the
ski slopes one sees these coloured molecules.

Among the thousands of different oil molecules, two can be pulled
out of a beaker to make a nylon rope. Do not ever assume that 'these days
everybody has seen this reaction!' For my adult audiences of non-scientists this is 'the best'. Joining top layer and bottom layer molecules alternately to make nylon before their eyes is chemistry coming to life. 'Do
something for me tonight—look at all the labels on your clothes—are
they made of wool or cotton, nylon (polyamide is Marks and Spencers'
name) acrylic, courtelle, terelyene or polyester? "The list of synthetics"
is long.—I have a "fleece" i.e. outdoor thermal clothing, it is polypropy-
lene made totally from oil molecules. No oil means no soap, no
medicines, no sports equipment, no make-up, no sun-tan cream. Solvents, plastics you tend to think of as nasty—but shuttlecocks, tennis
balls, scuba wear, surgeons' gloves, disposable sterile hospital equipment, computer chips all need solvents and plastics. All these everyday
things help make life as we know it and all are made from these fantastic
oil molecules.'

'Some of these fantastic molecules can take in UV light then they lose
some energy by wriggling about (i.e. vibrating!) and then the energy
released is visible light. Brighteners in washing powders do this so the
clothes literally glow with extra blue light and look "whiter than white".

You can see molecules fluorescing in washing powder, tonic, fluorescein and in certain rocks.'

More exciting than this is not re-emission of light but molecules in the dark releasing their own light like glow worms, fireflies and fish (Plate 36). We show the sort of reaction glow worms do, in a flask gleaming with turquoise blue light—there is no heat, just light from the luminol. 'When you see a reaction like that it makes you ask why is the light blue? Why does it have to be shaken? How long will it last? What are the actual molecules that are producing the light?' We tell the audience that 'if you take some TCP, some rhubarb and hair bleach you could get them to do this reaction (if you really understand the molecules).' (The bis(trichlorophenyl) oxalate ester decomposed with hydrogen peroxide in the presence of standard fluorescers.) Three tubes start glowing blue, violet and yellow and light up the darkness. They can last 12 hours and this reaction can be really useful. The reaction can be sold in plastic tubes labelled 'safety light'—12 hours. 'If you were in a dark place where a flame or electric light switch could cause an explosion or you were lost on the mountains or at sea, then you could start the reaction by bending the container to crack the inner tube—the molecules would mix and the safety-light would glow.'

We hope they have learned that molecules are fantastic things and some of the children will be enthused and inspired to learn about them. Plate 37 is a ten year old's recollections of his first encounter with chemistry. If our children are not enthused and excited we may lose them as future scientists and maybe the chance of another Faraday or Porter.

The children often ask me what was the best reaction I have ever done? I try to convey that fantastic moment when scientists know they have discovered something. I say 'you are holding the molecules in your hands and you know you are the first person to make them or find them, they may be the only sample in the Universe and that feeling of discovery is the moment you will remember all of your life.'

The reward for me is their wonder and enthusiasm. One little girl went home absolutely thrilled because she was made of stardust.

Some children will grow up to appreciate what Carl Sagan wrote. 'The earth is a place. It is by no means the only place. It is not even a typical place. No planet or star or galaxy can be typical because the cosmos is mostly empty. The only typical place is the vast cold universal vacuum the everlasting night of intergalactic space, a place so strange and desolate that by comparison planets and stars and galaxies seem achingly rare and lovely. If we were randomly inserted into the cosmos the chance that we would be on or near a planet would be one in a billion trillion trillion, 10^{33}, that is one followed by thirty three zeros. In everyday life such odds are called compelling. Worlds are precious. Our world is infinitely precious.'

I believe the care that comes from understanding its molecules is called Chemistry and that is why the lecture is called 'To Planet Earth with love from chemistry'.

References

1. Sagan, Carl, *Cosmos*.
2. Tilden, W. *Famous Chemists and their Works*. Routledge.
3. Holloway, J, private communication.
4. Sagan, Carl, *Cosmos*.

ANN HUBBARD

Born 1944, graduated from the University of Manchester Institute of Science and Technology and there obtained a Ph.D. in fluorine, mercury and silicon chemistry. Following this she worked as a research chemist in the Special Patent Unit of the pharmaceutical company SmithKline Beecham (then Beecham Research Laboratories). She moved into education in 1974 and, four years later, became Head of Chemistry at Reigate Sixth Form College where the Chemistry Department has grown from 40 to 120 'A' level chemistry students. She is the holder of the 1991 Royal Society of Chemistry Award for Chemical Education and has just been awarded the Salters' Chemistry Prize. She is currently the Public Lecturer for the Royal Society of Chemistry. Her assistant, John Hanson, was a research chemist at SmithKline Beecham.

Will computers ever permanently dethrone the human chess champions?

RAYMOND KEENE

Why are mind games, and chess in particular, important? Throughout the history of culture, prowess at mind games has been associated with intelligence in general. Goethe, the great German genius and polymath, once described chess as 'the touchstone of the intellect'. Human beings like to consider themselves as the most intelligent beings on the planet, but that position is now under threat from computers that can aspire to calculate up to one billion different chess positions every second. Indeed, last year in London the seemingly invincible world chess champion, Garry Kasparov himself, was sensationally crushed by a chess computer. I was present at the time and can testify to the shock waves that emanated around the chess world at the champion's humiliation.

However, I have some good news. Despite the awesome power of the computers, the human mind itself is equally awesome and is essentially capable of infinite connections and associations. The best human players are actually capable, on occasion, of playing perfect chess. Philosophically speaking the game of chess at the outset is a draw. In order to win at the highest levels one side or the other must take a risk to unbalance the position. This uncertainty principle brings with it exposure to loss. In my view, in the future, whatever level chess computers reach, championship matches between humans and computers will be open and level, with both sides scoring, so long as the humans remember that the basic barrier of the draw is an element working in their favour against the colossal calculating power of the machines.

The sole precondition for this mental arms race between humans and computers to be balanced is that chess should still continue to attract young players, fresh blood, who regard chess as a worthwhile career. If chess fails to do this the machines will take over completely.

This has already happened in draughts where the best player in the world is the computer program Chinook. Young players are no longer taking up draughts as a career. In comparison, chess as an international competitive sport seems healthy. The prize funds for championships are in millions of dollars as opposed to the few thousands available for draughts.

In my opinion humans will never lose consistently to computers so long as we collectively maintain the will, desire and commitment to beat them. Instead there will be a mutually beneficial and challenging mental arms race, a symbiotic relationship with both sides learning rapidly from the other, leading to games of hitherto unimagined depth and complexity.

In 1988, the first Computer Olympiad was held in London. Eighty-six programs from around the world competed for medals in games as diverse as Chess, Connect 4, Go, Backgammon and Bridge. In his concluding address, Professor George Steiner issued a grisly warning, namely that, with the march of computers, humans might be supplanted as the best players in some games, while other games would be solved and exhausted by computers. Thus such games would be peremptorily abolished as a source of human delight and intellectual challenge. No sooner had Professor Steiner concluded his discourse than it was announced that the Olympiad's Connect 4 tournament was not only the first, but also the last, which would ever be held. The winner, a Dutch program called Victor, had solved the game! Worse was to follow. At the end of August 1994, a £100 chess program on sale in the high street, working on a PC, defeated the World Chess Champion, Garry Kasparov, in a tournament game in London. This sensational result made world headlines.

The case for the computer

Richard Lang is an unrecognized British genius, 38 years old, married with two girls and one boy and now living in Poole, Dorset. Lang's modest, unassuming and even retiring manner belies the fact that he is one of the sharpest brains working at the forefront of artificial intelligence. It was his computer program, Chess Genius, with its ability to calculate 3.6 million different chess positions every minute that, on Wednesday, 31 August 1994, rocked the chess world by eliminating the world chess champion, Garry Kasparov, from the $160 000 Intel Grand Prix, in London.

'Before we beat Kasparov I though it would be ten to twelve years before computers could compete on level terms with the human champions' said Lang. 'However, after beating Kasparov I would now say that this time will reduce dramatically. One day machines will win all the time and take the world title. It's inevitable.'

Lang's program, Chess Genius, retails in disk form for less than £100 and runs on any IBM-compatible machine and is especially formidable on the Pentium processor. It was this combination that Kasparov faced in this fateful game. Chess Genius has won the world championship for microcomputers on nine occasions, the last of these being at Munich in 1993. This makes Chess Genius the reigning computer champion, just as Kasparov is the reigning human champion. Lang's profession is computer programming for commercial chess products, but as a student his subject was physics at London's Imperial College. He regards himself as near the lower echelons of international recognition in terms of his own chessplaying strength. 'My British Chess Federation rating would be about 180 which translates to approximately 2070 on the international scale,' said Lang. 'Nevertheless, although I do not play in tournaments, my actual understanding of chess is much greater, in terms of strategy and general rules.'

Herein, perhaps, lies the secret of Lang's spectacular and historic triumph against Kasparov. He actually tried to transfer his own intelligence into his machine, and when discussing his invention he embraces remarkably anthropomorphic terminology. 'Chess Genius does not adopt a brute force method, analysing all moves possible to immense depth. Instead my program uses its intelligence to throw out those lines which it perceives as not promising and searches more deeply into the lines it considers interesting. It has a lot of chess knowledge. It knows a great deal about the game.'

During August 1994 in Boston, Massachusetts, a Canadian computer program, Chinook, won the world championship in draughts, replacing the human champions for the first time. The number of different positions in draughts appears at first sight to be astronomical, namely 500 995 484 682 338 671 693. Chinook, however, was able to excel in draughts precisely because of its brute force approach. That number, colossal though it might appear, is infinitesimally small compared with the possible number of chess positions, which is approximately that number against but, staggeringly, multiplied by itself. In other words 10^{20} for draughts as opposed to 10^{40} for chess. Sheer brute force stands no chance in such a forest of numbers and this is where Lang's approach, using what he calls the machine's 'intelligence' to thread its way selectively through the tangle of variations, has evidently paid off.

Lang's partner is the 40 year old Munich architect Ossie Weiner. Having designed a Munich school he lost interest in his initial profession and switched to the testing and selling of chess computers. 'I provide the opening variations for Chess Genius and I also test it continually against all of the best of our rival programs,' Weiner said. 'Chess Genius is playing other programs all the time, day and night. They are all connected by cables. I look for mistakes, analyse the games carefully and constantly feed through my conclusions to Richard in England.'

Herr Weiner operates the machine in play, since he is an experienced tournament player himself and can make the moves more quickly, while entering them into the program's memory. His reaction to his team's victory against the world champion was, though, mixed. 'I have been having a repeated dream for years that I would sit across the board from the world champion with our machine and beat him face to face. When it actually happened, though, I felt so pitiful for Kasparov. It was a terrible moment. He was so shocked. I almost felt our victory had been hollow.'

Kasparov, who regards himself as being in the forefront of human resistance against the relentless advances of the machines, was clearly devastated by this defeat. During the two game match it was not even necessary to follow events on the board to discern what was happening—Kasparov's body language said it all. He spent long periods of time fidgeting uncomfortably, mumbling to himself and shaking his head in disbelief. Following his sensational elimination by the computer, he left the tournament hall immediately, without saying a word to the press.

Chess programs have caused upsets before, but these have invariably been in chess games played at very fast time limits—typically five minutes per game. At this sort of speed, even world class grandmasters can make serious errors and upsets are common. However, the Intel Grand Prix tournaments are held at the speed limit of 25 minutes per player, which is sufficient to avoid outright blunders and usually allows the better player to win.

Kasparov–Pentium Genius, London Intel Grand Prix, 1994, Slav Defence.

1 c4 c6, 2 d4 d5, 3 Nf3 Nf6, 4 Qc2 dxc4, 5 Qxc4 Bf5, 6 Nc3 Nbd7, 7 g3 e6, 8 Bg2 Be7, 9 0–0 0–0, 10 e3 Ne4, 11 Qe2 Qb6, 12 Rd1 Rad8, 13 Ne1 Ndf6, 14 Nxe4 Nxe4, 15 f3 Nd6, 16 a4 Qb3, 17 e4 Bg6. Kasparov has obtained a small advantage from the opening. He has good control of the centre and the black bishop on g6 is not participating in the struggle. **18 Rd3 Qb4, 19 b3 Nc8, 20 Nc2 Qb6, 21 Bf4.** In view of Kasparov's next move, 21 Be3 suggests itself here. **21 ... c5, 22 Be3 cxd4, 23 Nxd4 Bc5, 24 Rad1 e5, 25 Nc2 Rxd3, 26 Qxd3 Ne7, 27 b4 Bxe3+, 28 Qxe3 Rd8.** This is an excellent move, and one which may have been overlooked by the world champion. If instead 28 ... Qxe3+, 29 Nxe3 then White's control of the d-file gives him a substantial advantage. **29 Rxd8+ Qxd8, 30 Bf1**. 30 Qxa7 is impossible on account of 30 ... Qd1+ picking up the loose knight on c2. **30 ... b6, 31 Qc3 f6, 32 Bc4+ Bf7, 33 Ne3**.

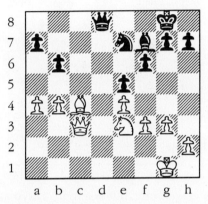

This harmless looking move is, in fact, a serious slip by Kasparov. The simple 33 Bxf7+ would have guaranteed an easy draw, but the next move allows the black queen to assume a dominating post in the centre of the board. **33 ... Qd4, 34 Bxf7+ Kxf7, 35 Qb3+**. The endgame after 35 Qxd4 exd4 is difficult for White, as his queenside pawns are exposed and the passed black d-pawn is a menace. However, in trying to avoid this, Kasparov gets his pieces completely tangled. **35 ... Kf8**. White's last chance now is 36 Kf1! and if 36 ... Qd2? 37 Nc4 Qxh2, 38 Nd6! Qh5, 39 Kg2 when Black is paralysed. If Black avoids 36 ... Qd2 the position is a draw. **36 Kg2 Qd2+, 37 Kh3 Qe2, 38 Ng2**.

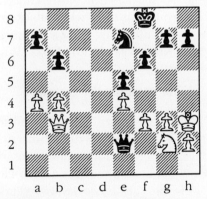

The White position has suddenly become very uncomfortable. **38 ... h5, 39 Qe3 Qc4**. This is an excellent manoeuvre. By combining threats on both sides of the board, the Pentium Genius creates fatal weaknesses.

This is a position where accurate calculation for a small number of moves can turn a good position into a winning one. This is, of course, the machine's forte and it has little difficulty in effecting the transformation. **40 Qd2 Qe6+, 41 g4 hxg4+, 42 fxg4 Qc4, 43 Qe1 Qb3+, 44 Ne3.**

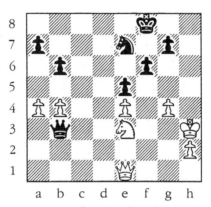

44 ... Qd3. This is very strong, but is a highly atypical computer move. Most machines, being ardent materialists, would have grabbed White's a-pawn with 44 ... Qxa4. This would not be a bad continuation but the white e-pawn is a greater prize than the a-pawn and the Pentium Genius has calculated that it is indefensible. **45 Kg3 Qxe4, 46 Qd2 Qf4+, 47 Kg2 Qd4, 48 Qxd4 exd4, 49 Nc4 Nc6, 50 b5 Ne5, 51 Nd6.** Kasparov must have been hoping for counterplay based on his threat of Nc8. Many human opponents could easily become nervous at this point and make errors. The Pentium Genius, however, is not flustered and calmly calculates its way to victory. **51 ... d3, 52 Kf2 Nxg4+, 53 Ke1 Nxh2, 54 Kd2 Nf3+, 55 Kxd3 Ke7, 56 Nf5+.** This is hopeless, but the problem with 56 Nc8+ is that after 56 ... Kd7, 57 Nxa7 Ne5+, the white knight is entombed on a7. The only way out is by bringing it to c6 when Black exchanges it off leading to a trivial win in the ensuing king and pawn endgame. **56 ... Kf7, 57 Ke4 Nd2+, 58 Kd5 g5, 59 Nd6+ Kg6, 60 Kd4 Nb3+, White resigns.**

The case for the humans

So, does this result herald the beginning of the end for human superiority in the field of chess? Many commentators think so. Frederic Friedel, a German computer chess expert, predicts that a computer will be world champion by the year 2000. Lang himself is a little more circumspect: he previously thought it would be ten to twelve years before computers could compete on level terms with the human champions but, following this recent success, he now thinks that this time frame will reduce dramatically. Even many chessplayers think the game is up, complaining that as the processors become yet faster the chess programs will soon become invincible.

But I don't believe it! At the moment, there exists a kind of collective fear amongst the top grandmasters when they face computer opposition. Knowing that small tactical inaccuracies will be ruthlessly exploited and that their opponent will never get nervous or tired, they approach such encounters like rabbits transfixed in the headlights of an oncoming car. I think that this is the main cause of Chess Genius's recent success. Competing against a computer is a completely different proposition from facing a human opponent but, at the moment, I have not seen much evidence of players adjusting their strategy to cope with the differing circumstances. Once the top grandmasters make a concerted effort to study computer chess programs, in the same way that they normally prepare for any serious opponent, then I think the results will be very different. Kasparov is a very proud man and I suspect he simply did not believe that he could be beaten by a mere machine. Next time he will have more respect and be better prepared.

The Pentium Genius did, in fact, register a further upset in the Intel tournament when it defeated Grandmaster Predrag Nikolic in the quarter-final stage. However, in the next round it came up against the Indian Grandmaster Viswanathan Anand and the two games played in this match demonstrated why computer programs still have a very long way to go before they will be able to score consistently against the top players.

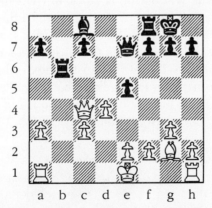

Pentium Genius–Viswanathan Anand, Intel Grand Prix, London, September 1994, Bogo-Indian Defence.
1 d4 Nf6, 2 Nf3 e6, 3 c4 Bb4+. Aware of the risks of engaging in hand to hand combat with such a tactically sharp opponent, Anand chooses a quiet opening variation. **4 Bd2 Qe7, 5 g3 Nc6, 6 Nc3 0–0, 7 Bg2 d5, 8 a3 Bxc3, 9 Bxc3 dxc4.** This move, surrendering Black's strong point in the centre, is unusual. Black normally continues with 9 . . . Ne4, aiming to

exchange off the white bishop on c3, or 9 ... Rd8, consolidating in the centre. **10 Ne5 Nd5.** This effectively commits Black to the sacrifice of a pawn but Anand judges that his central initiative will compensate. **11 Nxc6 bxc6, 12 Qa4 Nxc3, 13 bxc3 e5, 14 Qxc6 Rb8, 15 Qxc4 Rb6.**

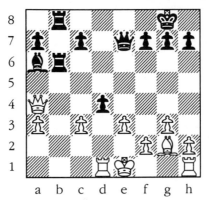

Although White is a pawn up, the best continuation is to return it with 16 0–0 Ba6, 17 Qa2 exd4, 18 cxd4 Rd8 (not 18 ... Qxe2, 19 Qxe2 Bxe2, 20 Rfc1, when White has a definite advantage in the endgame) and now White must lose either the d-pawn or the e-pawn and the position is equal. **16 e3.** This is a truly dreadful move and is a good demonstration of the limitations of chess-playing programs. No human player above the standard of moderate club player would even consider this continuation. After Black's reply, the white king is stuck in the centre and the black pieces have easy access into the white position. To the human eye, it is obvious that within a few moves White will suffer some horrible disaster. The problem for the computer is that it has no such intuition and, having calculated that it will not lose in the next three or four moves, sees no reason not to hang on to the extra material. The ability to calculate perfectly in the short term is of limited use if it cannot be allied to sound long term judgement. **16 ... Ba6, 17 Qa4 Rfb8, 18 Rd1 exd4.**

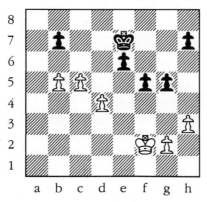

19 Qxd4. If 19 cxd4 instead, Black can win with the surprising and beautiful continuation, 19 ... Re6, threatening ... Rxe3+. **19 ... Rd6, 20 Qa4 Qf6**. Suddenly it is all over. If now 21 Rc1, Black wins with the attractive 21 ... Qxc3, 22 Rxc3 Rb1+ mating. Or 21 Qc2 Rxd1+, 22 Kxd1 Qd6+, and now 23 Kc1 or Ke1 allows 23 ... Bd3, while 23 Qd2 runs into 23 ... Rb1+, 24 Kc2 Rb2+ and the white queen goes. **21 Rxd6 Qxc3+, 22 Rd2 Rb1+, 23 Qd1 Rxd1+, 24 Kxd1 Bd3, 25 Ke1 Qxa3, 26 Bf3 g6**. Although the material situation is not too bad for White, his forces are hopelessly uncoordinated and cannot offer any resistance to the advance of Black's queenside pawns. **27 Be2 Bf5**. It is important for Black to preserve his powerful bishop. The Pentium Genius now kicks and struggles, but the result is not in doubt. **28 h4 Qc3, 29 h5 g5, 30 h6 Kf8, 31 Bf3 c5, 32 Rh5 f6, 33 Rh2 c4, 34 g4 Bd3, 35 Be2 Bb1, 36 Bf3 a5, 37 Bd1 Qb4, 38 f3 a4, 39 Rhf2 a3, 40 Kf1 c3, 41 Rd8+ Ke7, 42 Ra8 c2, 43 Ra7+ Ke6, 44 Ra6+ Kf7, 45 Ra7+ Ke6, 46 Ra6+ Kd5, 47 Bxc2 Bxc2, 48 Kg2 Bb1, 49 Rxf6 Qa5, 50 e4+ Kd4, 51 Rf5 Qa6, 52 Rxg5 a2, 53 Rxa2 Qxa2+, 54 Kg3 Ke3, White resigns**.

And now to show the further limitations of the computer's powers, I switch to one of its games against the Indian Grandmaster Anand from the semi-final. In this position, with Black to move, a simple draw can be achieved if Black plays 35 ... h6. The point is to present a defensive wall against penetration by the white king. Instead, remarkably, the computer blundered.

Viswanathan Anand–Pentium Genius, Intel Grand Prix, London, September 1994.
35 ... h5?? This error encourages White to punch a hole in Black's pawns. **36 h4! gxh4**. Black's case is suddenly hopeless. If instead 36 ... g4, 37 Ke3 followed by Kf4 and Kg5 simply mops up Black's pawns. Black can never return with his king to the kingside to defend against this invasion since after a move such as, for example, 36 ... g4 there follows 37 Ke3 Kf6, 38 c6 bxc6, 39 b6 and White immediately promotes a pawn. **37 Kf3 Ke8, 38 Kf4 Kd7, 39 Kg5 Kd8, 40 Kxh4 Kd7, 41 Kxh5, Black resigns**.

Fritz3/Pentium almost wins international master norm

For the first time in history a machine nearly achieved an international master norm in a regular grandmaster tournament, held in Bonn Bad Godesberg. The program Fritz3 running on an Olivetti 90 MHz Pentium PC scored 50 per cent to achieve a performance rating of ELO 2452. This puts it in the bracket of the top 50 players in Germany. The performance was more than 100 points better than that achieved by the IBM research computer Deep Thought in Hannover 1991, where it played in a grandmaster tournament and was rated 2326. Deep Thought costs millions of dollars, whereas Fritz3 is available in high street stores for less than DM, 200.

The most remarkable aspect of the Fritz3 Pentium performance is that it was not achieved in blitz or rapid chess but at regular tournament speed (two hours for forty moves). These time controls are supposed to favour humans, so grandmasters were confident of beating computers when they had so much time to ponder their moves. In Godesberg Fritz, in fact, scored a plus against the four grandmaster opponents with an incredible performance of 2652 against them. *The overall result was not so good because, ironically, the weaker international masters had prepared for the computer, using identical copies of the program, which can be bought in any chess store!* In other words, persistence, training, determination and sheer study, will pay off against both human and silicon opponents!

As an amusing conclusion to this argument let me quote an extraordinary case reported in *New in Chess* magazine issue 8, from 1994: '12 year old Tommy Friedel (German computer expert Frederic Friedel's son) played two games against an Olivetti M6 suprema computer with a 90 MHz Pentium processor running the Chess Genius program, essentially the same opponent that defeated Garry Kasparov. Tommy decided to copy grandmaster Anand's moves in his two wins against the Pentium Genius. Amazingly, the computer obligingly complied and lost two identical games. A triumph of preparation!'

RAYMOND KEENE

Born 1948 in London, educated at Dulwich College and Trinity College, Cambridge, specializing in German literature. Has written over 80 books on chess, a world record in this area, and has appeared on radio and TV in the UK and USA. Chess correspondent of *The Times* and *The Spectator*, he masterminded the 1993 World Chess Championship between Kasparov and Short in London, and is co-founder of the Mind Sports Olympiad project. Can play and win over 100 simultaneous games of chess. Organized the 1994 World Draughts Championship in Boston where, for the first time in any Mind Sport, a computer programme won the world title.

England's Leonardo: Robert Hooke (1635–1703) and the art of experiment in Restoration England

ALLAN CHAPMAN

No portrait or contemporary visual likeness survives of Robert Hooke, though when the German antiquarian and scholar Zacharias von Uffenbach visited the Royal Society in 1710, he specifically mentioned being shown the portraits of 'Boyle and Hoock', which were said to be good likenesses. Though Boyle's portrait survives, we have no idea what has happened to that of Hooke. It is curious, furthermore, that when Richard Waller edited Hooke's *Posthumous Works* for the Royal Society in 1705 he did not have this picture engraved to form a frontispiece to the sumptuous folio volume.

On the other hand, we do possess two detailed pen-portraits of Hooke written by men who knew him well. The first was that recorded by his friend John Aubrey, and describes Hooke in middle life and at the height of his creative powers:

> He is but of midling stature, something crooked, pale faced, and his face but little below, but his head is lardge, his eie full and popping, and not quick; a grey eie. He haz a delicate head of haire, browne, and of an excellent moist curle. He is and ever was temperate and moderate in dyet, etc.

The second is that by Richard Waller, whose forthright account of the elderly Hooke can scarcely be said to err on the side of flattery:

> As to his Person he was but despicable, being very crooked, tho' I have heard from himself, and others, that he was strait till about 16 Years of Age when he first grew awry, by frequent

practicing, with a Turn-Lath ... He was always very pale and lean, and laterly nothing but Skin and Bone, with a Meagre Aspect, his Eyes grey and full, with a sharp ingenious Look whilst younger; his nose but thin, of a moderate height and length; his Mouth meanly wide, and upper lip thin; his Chin sharp, and Forehead large; his Head of a middle size. He wore his own Hair of a dark Brown colour, very long and hanging neglected over his Face uncut and lank ... [1]

The context of the New Science

Before examining Hooke's life and researches, however, it is important to look at their context, for his contributions to physical science came at the end of a period of a century and a half during which the once-coherent structures of classical science had received one blow after another. The explanations of the natural world which medieval European scholars had inherited from classical Greece had been static in character and based on a series of apparently self-evident principles. All changes of matter could be explained by the interaction of the four elements, Earth, Water, Air and Fire, as the principles of solidity, wetness, volatility and heat endlessly mixed and separated.

These four elements also lay at the foundation of all living things. The hearts of all living creatures generated a spontaneous, or innate, heat, that was radiated throughout the body by the blood, while the life-principle of air, or *pneuma*, both intermingled with the blood in respiration, and also helped to cool the heart. Heat rose, cold congealed, 'grass became flesh', and flesh decayed when its life principle had departed.

While this flux of elements prevailed on the earth, the heavens were made from a perfect, stable fifth element. Because they were made of one single and changeless substance, the stars and planets moved with a geometrical precision that nothing on earth could ever emulate, thus exemplifying that deep dichotomy between terrestrial and celestial that lay at the heart of all classical science.

The ancient Greeks, and most significantly Aristotle, had devised a complete taxonomy of nature based on these principles by 350 BC, and for the next 1900 years it proved capable of answering most of the questions that could be addressed to it. It was a magnificent intellectual achievement, though it embodied a conservative approach to knowledge, and like librarianship or museum curation, saw its first duty as absorbing, classifying and preserving the known rather than exploring pastures new.

But after 1492, the assaults on its all-encompassing explanatory credibility began to increase. The discovery of America fundamentally discredited ancient geography. Tycho Brahe's supernova of 1572 and Galileo's telescopic discoveries after 1609 similarly shook classical astronomy. Rapid developments in optics and mechanics, moreover, seemed to indicate that phenomena could be studied amidst the four chaotic elements of the Earth that were just as mathematically exact as those observed in the heavens, while in 1628 William Harvey discovered that the heart was a pump, and not a furnace. All of these discoveries flew in the face of the classical writings, and showed that the 'moderns' might well know more than the 'ancients'. None of these discoveries, moreover, were the fruits of speculative philosophies; they were physical discoveries. Passive observation could classify, but experiment could break into realms of new knowledge. In the words of Sir Francis Bacon, who more than anyone else championed the cause of experiment, and whose writings directly inspired Robert Hooke and the early Fellows of the Royal Society, nature must be 'put to the torture', and made to yield its reluctant secrets to the astute investigator. It was not for nothing that Bacon's distinguished legal career took place in one of the most sanguinary periods of English constitutional history!

And as the judicial inquisitor needed his special tools of assault and persuasion to make his victim speak, so the scientific experimentalist needed his, for the laboratory—which included the newly invented telescope, microscope, airpump, thermometer, and many other instruments that refined the perceptions—was the torture chamber wherein long-secretive nature would be cross-examined.

The radical reappraisal of how nature worked that was taking place in the early seventeenth century was also rich in perceived religious implications. Far from being persecuted by the Church, indeed, we must not forget that the Scientific Revolution was seen as fulfilling Old Testament prophecies. Hooke expressed the prophetic character of the New Science very succinctly in the Preface to *Micrographia* in 1665:

> And as at first, mankind fell by tasting of the forbidden Tree of Knowledge, so we, their Posterity, may be in part restor'd by the same way, not only by beholding and contemplating, but by tasting too those fruits of Natural Knowledge, that were never yet forbidden. [2]

In the spirit of Bacon's and the Royal Society's motto, *Nullius in Verba*, it was not only to be by passive word-exercises that mankind would reach a profounder understanding of the Divine Creation, but also by action and experiment.

Yet this new mastery of nature to which the age was laying claim had a darker (or more ecstatic) dimension, depending on one's perspective. After the Fall of Mankind as a result of excessive curiosity, as recounted in *Genesis*, humanity had been bounded within a fixed scheme of knowledge, though ancient prophecies had indicated that new enlightenment would come to man shortly before the end of the world. Many seventeenth-century scholars had computed from the prophetic books of the Bible that Armageddon was now at hand, and no prophecy fitted the age better than that from the Book of Daniel, XII. 4:

> Many shall run to and fro, and Knowledge shall be increased.

The geographical discoveries, the religious wars of the Reformation, numerous new inventions, supernovae, Jupiter's moons, the execution of King Charles I, the discovery of the microscopic realm and of the vacuum, and the refutation of the truths of Aristotle's science: all were clear fulfilments of Daniel's and many similar prophecies. The search for religious meaning lay at the heart of seventeenth-century intellectual culture, and to dismiss it from our understanding of their science produces a picture as lopsided as that which would result if a historian in 300 years' time wrote of the twentieth century in a way that dismissed the significance of economics.

Hooke's origins and early career

Robert Hooke was born on 18 July (Old Style) 1635 at Freshwater, Isle of Wight, where his father John was curate in charge of the parish. The Reverend John Hooke had fulfilled a variety of curacies on the island since at least 1610, and had been at Freshwater since 1626. As a boy Hooke was not strong, and his father, who was reluctant to subject him to the rigours of a boarding school, educated him at home. It was during these early years, however, that his talents began to manifest themselves. He was an extraordinarily quick learner, possessed a manual dexterity which enabled him to build an impressive array of mechanical devices, and his untrained draughtsmanship so struck the visiting artist John Hoskins that he advised Mr Hooke to settle upon an artistic career for Robert. In 1648, the Reverend John Hooke died, and while Robert was believed to have received a legacy of £100, the recent discovery of John Hooke's Will indicates that he only received £40, a wooden chest, and some supplementary payments [3].

Though we have no idea who made the arrangements, the thirteen-year-old Robert now went up to London for 'tryall' in the studio of Peter (later Sir Peter) Lely, the leading portrait painter of the age. According to the account that the middle-aged Robert Hooke gave to his biographer friend, John Aubrey, he 'quickly perceived what was to be done' in painting, and complaining that the oils and varnishes irritated his chest, left the studio to be enrolled at Westminster School under Dr Richard Busby. One presumes that the thirteen-year-old must have had friends, and a patron in London, for these adroit social moves would have been extraordinary, even for a youth of Hooke's precocity, unless he had received help.

At Westminster, Robert Hooke found his feet in the city that was to provide the theatre of operations for the greater part of his career. Dr Busby, who was still within the first decade of his fifty-five-year reign as Head Master of what he was turning into the most intellectually distinguished school of the seventeenth century, quickly recognized Hooke's genius. At Westminster, Hooke acquired a mastery of ancient languages, learned to play the organ, contrived flying machines and mastered the first six books of Euclid's *Elements* in a week. He remained on warm terms with Dr Busby for the rest of the Head Master's eighty-nine-year life, undertaking architectural design work for him and for Westminster Abbey, and mentioning him in his Diary.

In 1653, Robert Hooke left Westminster to take up a poor scholar's place at Christ Church, Oxford, where he was described as 'Servitor' to a Mr Goodman. He was also to have been a Singing Man in the Cathedral, though as the abolition of the Anglican Church between 1643 and 1660 would have closed down the liturgical choirs, one presumes that Hooke received a Singing Man's modest endowment by way of a scholarship. But Robert Hooke was clearly a man who possessed musical abilities.

It is not known whether Hooke was especially inspired or encouraged by John Owen, the new Dean and Oliver Cromwell's ex-Chaplain, who had been imposed on once-Royalist Christ Church by the Puritan Commissioners after the Civil War, but he fulfilled the necessary requirements to take his BA and MA degrees. However, Oxford inspired Hooke in other, non-curricular ways, for it was in the University city that he fell in with that group of men who within a decade would form the original Fellowship of the Royal Society. It was in Oxford that Robert Hooke began his apprenticeship to science and formed a collection of influential and creative friendships. Dr John Wilkins, Warden of Wadham College, who was the leader of the Oxford scientific 'Club', encouraged him in astronomy, mathematics, and mechanics, as did the young Christopher Wren. Hooke was employed as chemical assistant by the distinguished anatomist Dr Thomas Willis and from him, and from Richard Lower, he probably acquired those dissection skills which would be essential in his own later researches into respiration. But his own most important single contact in Oxford was with Robert Boyle, whose Assistant he became around 1658, after leaving the employment of Dr Willis. From Boyle, he gained a thorough mastery of chemistry and practical laboratory skills, while Hooke in his turn contributed his own talents as a mechanician to Boyle's own researches into air. But we will return to Boyle when looking at Hooke's experimental researches. [4]

As an intelligent and experienced 'operator' to Willis and Boyle, and as a man of acknowledged talent as a practical exponent of the 'new philosophy', Robert Hooke was the obvious choice for the office of 'Curator of Experiments' to the newly chartered Royal Society of London in 1662. But we must not forget that at this early stage in his career he did not sit with the Fellows—Boyle, Wren, Wilkins and others—as an equal, but as an employee (or 'servant' in seventeenth-century language) with a salary yet to be paid. Unlike them, he was not a '*Gentleman*, free, and unconfin'd' possessing independent means, and the original terms of his employment required him to produce demonstrations at Royal Society meetings, as well as receive 'orders' for the undertaking of particular research investigations. [5] In 1665, however, his academic standing became more regularized when he was appointed Professor of Geometry at Gresham College, with rooms and accommodation in the same building wherein the Royal Society held its meetings; by then he was already a full Fellow of the Royal Society. By 1665, therefore, Hooke had acquired those physical circumstances that would see him through the remaining thirty-eight years of his life. They also made him the first salaried research scientist in Britain. [6]

Hooke's scientific ideas

Mid-seventeenth-century Europe was a veritable market-place of competing philosophies of nature in the wake of the confusion that followed the eclipse of Aristotelianism. Though historians of science generally speak of the rise of the 'mechanical philosophy' at this period, one should remember that this is a portmanteau designation for several quite distinct 'systems' that shared the speculative premise that energy was transmitted by particulate collision. The most uncompromising of mechanists was Thomas Hobbes (better known today as a political philosopher) who argued that matter and the laws of motion could be made to explain everything, from celestial mechanics to the appearance of ghosts. René Descartes saw all physical, but not spiritual, phenomena as occasioned by an endlessly agitated aether, the vortices and swirls of which carried along the particles that produced physical motion. Pierre

Gassendi revived the once-called Godless doctrine of atomism, and conceived of matter in terms of the geometrical arrangement of fundamental particles guided by the hand of God. And especially popular in England were the ideas of Francis Bacon, which were concerned less with the innermost structures of matter in motion and more with developing the correct experimental method and arranging the results into taxonomic schemes.

Robert Boyle, in his chemical investigations, was drawn to a Christianized version of the atomic theory, where the geometrical arrangement of the atoms defined the chemical characteristics of the substance. And as a Baconian, Boyle devised meticulous courses of experiments by which he hoped to test these ideas. As Boyle's assistant, one can expect Robert Hooke to have been influenced by his master's ideas, though there are important points of divergence.

While it goes without saying that Hooke was an experimentalist in the Baconian tradition, it is obvious to anyone who reads Hooke's writings that he was no methodological purist. As every modern scientist now knows, no original investigator can be the rigid adherent of a pre-determined method, for creativity in science is more than recipe-following. Robert Boyle, and Robert Hooke, however, were probably the first scientists to encounter this fact of life, for while they were not the first men to perform experiments, they were the first to undertake whole courses of experiments and, in Hooke's case, conduct them in disciplines as diverse as physics and physiology. *Micrographia*, which published the results of a series of observations and experiments conducted between 1661 and 1664, should be required reading for every science undergraduate, for it amply demonstrates how brilliantly eclectic, yet how tightly controlled, a series of physical investigations can be. It showed how the microscopical examination of ice crystals could lead to a discussion of atomic structures; how the first recognition of the cellular structure of wood initiated research into the role of air in combustion; and how the anatomical description of a fly developed into an experimental essay in aerodynamics, acoustics, and wave-patterns.

In published researches covering nearly forty years, Hooke was constantly casting around for a consistent, underlying principle that could be shown to bind the whole of nature together: a 'Grand Unified Theory', as it were. That nature did contain common lucid principles would have been taken as axiomatic by Hooke, for as the entire universe was the product of divine intelligence, it was inconceivable that God could be inconsistent in His grand design. And as human intelligence was congruent with that of God, it stood to reason that the key should be within man's reach. As Kepler had said, science was thinking God's thoughts after Him.

Though Robert Hooke never came up with a Grand Unified Theory that could be made to stand experimentally in all cases, one can extract a series of principles which run as a thread through his thought. One of these was a version of the atomic theory of matter, though he was careful not to push it too far, for lack of clear experimental evidence. Yet Hooke's atomism is more dynamic and rooted in motion than that of his master. Boyle, as we have seen, held to a broadly geometrical concept of atomic arrangement, whereas Hooke's was more kinematic and based on energy, or pressure, constantly exciting a medium so that the atoms became the efficient causes of all things. [7]

And when one came to the medium, or aether, in which the atoms were suspended and through which they received their powers of impulse, Hooke seems to have held different ideas at different times, depending on the results of particular researchers. Was the aether itself a 'stagnant', passive agent through which atomic collisions took place (in the way that railway lines are passive agents down which colliding waggons move), or was it the aether that originated the motion? Hooke considered that the primary forces of nature, such as light, magnetism, and gravity, might act through aethers, or parts of the aether, that were peculiar to themselves. [8]

As a mechanist, Hooke needed a medium of some sort if a cause were to produce an effect, for without a physical connecting agent, no matter how tenuous, one was no better off than the magicians who happily explained cause and effect by means of occult sympathies. One fundamental way in which Hooke differs from modern (or post-Newtonian) scientists is in his concern with active principles and connecting mediums, for like most other seventeenth-century researchers, he was still a 'philosopher' who was interested in the causes of things. Though he, like Boyle, Descartes and Gassendi, had abandoned the Greek qualitative approach to nature, in favour of a mechanical, quantitative approach, his thought processes were still haunted by the sources of cause and effect, albeit re-dressed in mechanical garb. It took Newton and the scientists of the eighteenth century to bequeath causes to the metaphysicians, and concentrate on expressing the nature of effects in precise mathematical terms.

If there were one single mechanical principle which Hooke saw as present in most parts of physical nature, it was vibration. In many branches of research, he saw vibration as the thing which moved from an active source, through its appropriate aether, to produce a measurable effect. We will return to Hooke and vibration when looking at his work on spring and the elasticity of bodies.

Though Hooke might have been happy to entertain the presence and characters of atoms and aether when speculating about an ultimate metaphysic for science, he had a clear understanding of what had made the scientific discoveries of the age possible: an enhanced ability to perceive and quantify nature by means of instruments. It is in the long Preface to *Micrographia* in 1665 that he sets out most clearly his scientific manifesto and speaks of instruments as devices which lend new investigative power to relatively imprecise human sense-perception:

> The next care to be taken, in respect of the Senses, is a supplying of their infirmities with Instruments, and, as it were, the adding of artificial Organs to the natural. [9]

In his stress on the primacy of the senses in all perceptions of nature, from everyday experience to sophisticated research, Robert Hooke became one of the founders of the British empirical tradition and an influence on figures like John Locke. Hooke, moreover, did not merely talk about sense-knowledge, but made it the very king-pin of his experimental technique, realizing that, if one were going to investigate the 'animalcules' in water, the surface of the moon, or the vacuum, then the senses would need artificial enhancement by means of instruments. The invention and use of instruments, indeed, runs through his entire career, from his first devising of an airpump for Boyle in 1659 to his last recorded scientific utterance in December 1702, when, according to the *Posthumous Works* (p. xxvi), he tried to devise an improved instrument to measure the horizontal solar diameter, 'but discovers not the way'. It is true that Aristotle had placed an emphasis on the senses when examining natural phenomena, but to him the reality of a thing was defined in the totality of its parts as perceived by the gross senses. But what the new, instrument-based, experimentalists introduced into sensory perception were new options whereby a scientific reality could be defined. Was the correct definition of a horse that of a large quadruped, or was it the mechanics of its skeleton and muscles, its heart-rate related to body weight, or particular characteristics of cells and blood as seen under the microscope?

Having looked at the principles that underlay Hooke's scientific thought and approach to knowledge, it is now time to examine some of his researches in more detail.

Robert Hooke's researches

1. Air and combustion

For two thousand years up to the seventeenth century, air had been regarded as a stable, simple, vital force of nature, and one of the four elements. According to Aristotle, it must invade all spaces not occupied by one of the other three elements, for the universe was full and intact, with no unoccupied interstices. Nature, therefore, abhorred the concept of the vacuum, for in a balanced universe, actual vacuums could not exist. It was seriously disconcerting for the loyal adherents of Aristotle's physics after 1643, therefore, when Galileo's pupil, Evangelista Torricelli, found an empty space in the sealed glass tube, above the mercury, in his newly invented barometer. Philosophers across Europe tried to devise ways of establishing the properties of the 'Torricellian vacuum', and the anti-Cartesian Frenchman, Giles Persone de Roberval, had the idea of inserting a deflated and sealed carp's bladder into the barometer tube before it was filled with mercury and inverted. The apparently deflated bladder suddenly inflated in the vacuum, as the residual air inside it expanded to fill the vacant space. [10]

To investigate this new, instrument-generated phenomenon of nature, however, it was necessary to make a vacuum that was physically larger and more accessible than that inside a barometer tube. Otto von Guericke in Germany pumped air out of a sealed barrel, invented the famous 'Magdeburg Hemispheres' which needed teams of horses to separate when evacuated, and showed that air pressure bore down intensely upon an evacuated space.

It was when Robert Boyle began his own researches into air and vacuum around 1658 that Robert Hooke made his début on the scientific stage, for while he had previously worked for Thomas Willis in Oxford, it was with Boyle that his creative scientific genius first found expression. Originally, Boyle had gone to Ralph Greatorex for the construction of an airpump which was to be more versatile than von Guericke's, but it had been a failure. How Hooke, as an entirely academically trained individual, without any formally imparted craft skills behind him, had built or perfected a successful pump is hard to explain, for Greatorex was the leading pumping engineer in England, and a man who had made a considerable amount of money in draining the Fens. [11] But Boyle gave Hooke acknowledgement in his published researches, though it is clear that even with Hooke's machine, vacuums were difficult to maintain, and pistons, cylinders and valves were liberally coated with 'Sallad Oyl' to make better seals. It took about three minutes of hard pumping to get a good vacuum, and some additional action was probably necessary if it were to be kept for several minutes, as the air squeaked and whistled through the imperfect seals. Hooke does not seem to have done the manual work during the experiments, for Boyle refers to 'our pumper' as a third party. [12]

Apart from any new features that were special to the pumps or valves,

Hooke's machine contained three design features that were of the greatest significance. The first of these was a large glass vessel some fifteen inches in diameter, called the 'Receiver', as the space to be evacuated. Secondly, a brass stopper some four inches in diameter set into the top of the receiver made it possible to gain easy access to the experimental area, and seal everything up before the pumper set to work. Thirdly, an ingenious secondary brass stopper with conical sides passed through the large stopper, so that when liberally coated with salad oil, it could be turned around without breaking the air seal. This rotating stopper could be used to pull a thread to actuate some experiment *in vacuo*. With Hooke's machine, therefore, the experimenter had easy physical access to a fairly large experimental site that was entirely visible through the glass receiver. It was to be used to conduct a series of experiments which needed clear vision and the ability to ignite and move things. [13]

Candles, glowing coals, and slow-match all went out in the evacuated receiver, though the coals could be revived to glow again spontaneously if air were re-admitted in time. Boyle and Hooke were especially interested in the behaviour of smoke that rose from the extinguished candle wicks. Did some all-pervading aether that was much more subtle than air still bear the smoke upwards before it touched the inner walls of the receiver and descended? Was the Aristotelian doctrine true in so far that all fire-products rose upwards—even *in vacuo*? But it was impossible to be certain about the true cause of the smoke's behaviour, for both Boyle and Hooke realized that they did not have perfect vacuums, and a very small quantity of residual air was always likely to be present. [14]

What the airpump experiments did demonstrate beyond all doubt, however, is that the element air was greatly elastic and capable of much rarefaction and compression. Hooke found, for instance, that when a burning candle was placed inside a sealed vessel of a particular size, the flame went out in about three minutes. But when the air was compressed by pumping into the same vessel, it burned for fifteen minutes, thereby showing that the active principle in air which sustained a flame in any given volume could be altered mechanically. When Boyle repeated Roberval's carp bladder experiment with that of a lamb in the capacious receiver of the Hooke airpump, he concluded that air could expand by a factor of 152 to fill the vessel. [15]

But in many ways, Boyle's interests in air differed from those of Hooke, for while Boyle was primarily interested in the physical properties of aerial elasticity, and how it related to his ideas on atomic structures, his assistant's interests tended to be more chemical. In particular, they related to Hooke's interests in combustion and in what he conceived as a corrosive property in air.

Almost certainly, Hooke's interests in air and its role in combustion stemmed from Boyle's original researches into saltpetre, or nitre, as the seventeenth-century chemists called potassium nitrate. Boyle had conducted these experiments around 1655, several years before the airpump was built, as part of an investigation into the nature of saltpetre's 'inflammable principle' and its relation with air. A crucible of saltpetre, or nitre, had been heated until it melted. Pieces of charcoal were then dropped into the fused mass, and each piece immediately combusted into flames and smoke. After enough charcoal had been dropped in, however, no more combustion took place, and the washed residue was found to be inert. But when this residue was mixed with nitric acid, Boyle observed that the easily identified crystals of saltpetre began to 'shoot' in it once again, and when prepared and dried, could be heated so as to consume more charcoal. [16]

It seemed, indeed, that the so-called 'inflammable principle' could be driven off by burning, and then restored by a 'nitrous' chemical substance. Was burning, therefore, not an innate force of nature, but a chemical process where ingredients were exchanged between the air and chemical substances? If this were the case, it seriously challenged Aristotle's doctrine of combustion as caused by the element fire, and further implied that Aristotelian air was not a pure element, but possessed chemically reactive properties.

The challenge to Aristotelian ideas was strengthened when Thomas Willis, who was Hooke's first Oxford employer, began to experiment with a new preparation named *aurum fulminans*, or exploding gold (gold fulminate), which could explode without fire. All that one needed to do was to place a little *aurum fulminans* on to a spoon, and cover the spoon with a heavy coin. If one gently tapped the spoon on to a table top, the chemical exploded violently and blew the coin up to the ceiling. [17]

If *aurum fulminans* could create fire without a spark, it was shown during the winter of 1659–60 that gunpowder, which is rich in saltpetre, could be fired in the evacuated receiver of the airpump without the need for air. The low winter light and the irregularities in the glass made it difficult to ignite the gunpowder by means of a burning-glass focusing the sun's rays. Instead, Hooke devised a frame that could be secured inside the receiver upon which was fastened a cocked pocket pistol with a pinch of gunpowder in its flashpan. By turning the airpump's secondary brass stopper, with its well-oiled airtight seal, a piece of string could be used to pull the trigger of the pistol. The ensuing discharge of the priming powder *in vacuo* clearly suggested that air was not necessary for the discharge of gunpowder. [18]

By the early 1660s, Robert Hooke had developed a theory of combustion in which the elastic medium of air possessed two quite separate properties. It contained a 'nitrous' part, which was capable of reacting with substances to produce combustion or explosion, and a 'fixed' or inert part. He outlined the theory and the experiments by which he tried to substantiate it in 'Observation 16; Of Charcoal' (p. 103) in *Micrographia*. Hooke envisaged air as a powerful dissolving agent, or, in the nomenclature of the early chemists, a *menstruum*. Whenever 'sulphureous', or potentially inflammable, substances like wood were heated to a particular point, their atoms were furiously descended upon by the *menstruum* of 'nitrous air' or 'aerial nitre'. The potentially volatile 'sulphurs' within the substance were thereby released, and an inert ash remained. Commonly combustive materials like wood and candle wicks could not burn *in vacuo* because even when locally heated by a burning-glass no aerial nitre was present to dissolve them. Gunpowder exploded *in vacuo*, however, for it contained its own inbuilt supply of 'fixed' nitre, in its saltpetre.

In his Cutlerian Lecture *Lampas*, delivered to the Royal Society in 1677, Hooke developed his ideas on combustion when he analysed the parts of a lamp or candle flame. He noticed that the point of combustion appeared to be at the bottom part of the conical flame, where the oil rising up the wick became excited by the heat above it. At a critical point, it was devoured by the aerial nitre, and produced the tulip-shaped inner flame, where the rising sulphurous particles or atoms made contact with the aerial nitre to produce a glowing combustive interface. He also realized that the interior of the flame did not emit light, but only the tulip-shaped, combustive interface around it. The interior consisted of heated but non-luminous sooty particles that had failed to go off, as it were, and simply rose as greasy smoke. It was within this dark, sooty interior that the non-light-emitting part of the wick lay, and Hooke noticed that when this spent wick fell over, and broke through the combustive interface, it glowed red, as it entered the aerial nitre that surrounded the flame. [19]

Hooke obtained this information by inserting thin plates of glass and 'Muscovy glass' (mica) in the flame, both from above and bisecting the flame sideways, to reveal the light and dark zones of the interior. He also used powerful sunlight to project an image of a candle flame on to a whitewashed wall, whereby he could discern the dark interior and heat zones in the resulting shadow.

Hooke's researches into combustion and the airpump naturally led him to the physiology of respiration. In this immediately post-Harveian age, physiologists were investigating the relationship between blood circulation and respiration, and the deaths of birds and small animals

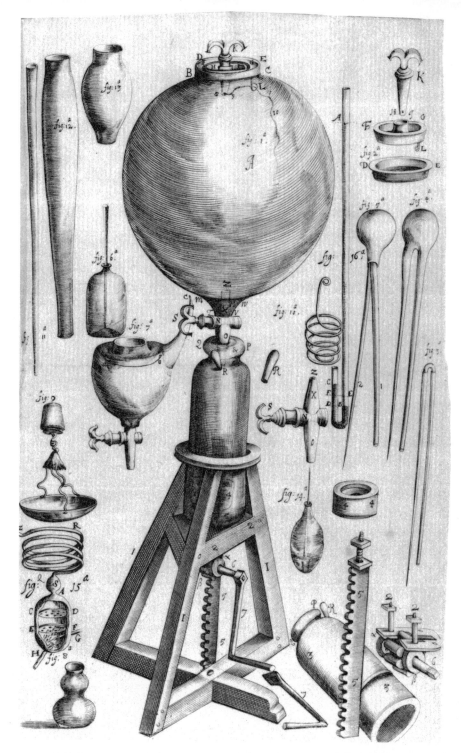

Fig. 1 Hooke's Air Pump built for Robert Boyle, 1659–60. From G. Keynes, A Bibliography of Dr. Robert Hooke, Clarendon Press, 1960.

that were placed in the receiver of the airpump demonstrated the importance of 'vital' air to life in a way that we take for granted today. The circulatory physiology of William Harvey recognized that the lighter colour of arterial, as opposed to venous, blood had something to do with its having picked up air in the lungs. In 1667, Hooke suggested that the blood might pick up 'aerial nitre' in the lungs. Boyle noticed that fresh lamb's blood inside the airpump receiver frothed as the pressure fell, and John Mayow repeated and refined the experiment. When blood from the veins was put into the airpump, however, nothing happened. [20]

The asphyxiating properties of the newly discovered vacuum were sufficiently impressive that when a visiting dignitary from Denmark was entertained by the Royal Society, around 1663, an anonymous wit wrote the following verse:

> To the Danish Agent late was showne,
> That where noe Ayre is, theres noe breathe;
> A glass this secret did make knowne
> Where[in] a Catt was put to death.
> Out of the glass the Ayre being screwed,
> Pusse died, and ne're so much as mewed. [21]

To study the effect of low atmospheric pressure on a human being, Hooke devised a sealed chamber in 1671, out of which the air could be evacuated. He sat in the chamber himself, while his 'pumper' took the pressure down to well below normal atmospheric pressure. This courageous experiment, in which Hooke experienced pains in his ears and deafness, probably made him the first person to experience 'high altitude' sickness, and monitor its effects. [22]

Within a decade, the Aristotelian explanations for burning and breathing, and the plausibility of two of the four elements, had been fundamentally challenged. In the early 1670s, moreover, Dr John Mayow, of Wadham College, Oxford, had started to take Hooke's ideas one stage further, when he discovered that burning and breathing competed with each other for the same active ingredient in an otherwise inert volume of air. [23] But it would be historically incorrect to see Hooke with his aerial nitre as a proto-discoverer of oxygen, for seventeenth-century researchers had no real concept of chemically specific gases, and still couched their ideas in terms of 'vital principles' and what we might call allotropic states of air. Yet the very fact that air might have allotropic states, that it was vastly elastic, and that it might somehow be able to fix its vital parts in stable chemical substances, signalled a fundamental shift in ideas about the natural world. And all these new ideas had been gained not by speculation, but by 'putting nature to the torture' in a carefully planned enquiry that hinged on a newly invented piece of apparatus.

2. Microscopy

The microscope was invented some thirty years before Robert Hooke was born. The Yorkshire scientist Henry Power had published microscopical observations before Hooke, and in 1661, Marcello Malphigi had used the instrument to provide clinching evidence in favour of Harvey's theory of blood circulation when he discovered the capillary vessels in the lungs of a frog. Yet for over half a century after its invention, the microscope had been a poor relation to the telescope in terms of its ability to produce fundamental scientific discoveries. Not until Robert Hooke published his own microscopical researches, in 1665, was it made manifest to the scientific world that the microscope revealed an organized realm of nature that was as diverse in its structures and as vast in its scale as the telescopic universe. For centuries, indeed, and long before the invention of the telescope, philosophers had speculated about the vastness of space, though no one had thought seriously about the existence of living creatures that were smaller than cheese-mites or inanimate objects smaller than dust particles. It is true that the atomists had conjectured about the existence of the minuscule particles that composed matter, but these had been objects of a philosophical character, which held out no hope of physical detection.

When Hooke's *Micrographia* first appeared in the bookshops in January 1665 at a lavish thirty shillings per copy, therefore, it had a quite sensational impact. It bowled Samuel Pepys right over and transfixed him in his chair until two o'clock in the morning; 'the most ingenious booke that ever I read in my life'. [24] More than anything else, it whetted Pepys's appetite for the New Science. He subsequently bought instruments, joined the Royal Society in February 1665, and in 1684 became its President. *Micrographia* was one of the formative books of the modern world, and like all influential pieces of writing, was capable of triggering responses on many different levels of understanding.

Within the scientific community, it provided one of the most articulate and beautifully presented justifications for experimental science ever produced. Mere observation, after all, could take one no further than Aristotle had gone in his descriptions of animals and natural forces, but when observation was refined by means of specially designed instruments, and used to 'put nature to the torture' in a context of addressed questions, then remarkable discoveries could be made. *Micrographia* not only provided a wealth of new data for science to consider, but showed how experimental investigations could be built upon them. A seemingly simple observation of a piece of charcoal under the microscope, for instance, could lead to a recognition of the presence of cells, to an investigation into burning, and to Hooke's work on the dissolving properties of aerial nitre. None of the Observations in *Micrographia* are simple; all of them are detailed starting-points for further physical investigations in one way or another. Hooke showed that sense knowledge could be reliable when used within the correct disciplinary restraints, and what the body could physically perceive via its 'artificial Organs' left little doubt that the experimental method actually worked.

If *Micrographia* was so important within the scientific community, it must be remembered that its influence on the cultured laity, like Pepys, was equally profound. The book was written in an easy style that would have been accessible to any innumerate who could read Shakespeare or the Bible, for Hooke could write vivid and powerful prose. It was, moreover, the first proper picture-book of science to come off the presses, for its sixty Observations were accompanied by fifty-eight beautiful engravings of the objects seen beneath the microscope. Hooke's artistic gifts had been essential to *Micrographia*, for only a man who could faithfully interpret and delineate the awkward images that were produced by the compound microscopes of the 1660s could envisage such a book in the first place. Modern science is replete with visual images, and the televisual image is the most powerful medium through which its ideas are now communicated to the lay public. We must not forget that this tradition of visual communication largely begins with Hooke's *Micrographia*.

Part of the popular fascination of *Micrographia* lay in the arresting new perspective that it cast on to common and familiar objects: a fine needle point looked like a rough carrot (Observation 1, p. 1), delicate silk looked like basket-work (Observation 4, p. 6), and extinguished sparks resembled lumps of coal (Observation 8, p. 44). But it was the observations of moulds of various kinds, 'Eels in Vinegar', and of insects that were the most sensational (Observations 53 and 54, pp. 210–11). That a flea could be depicted with the anatomical precision of a rhinoceros was quite shocking, and one wonders how many nightmares were occasioned by *Micrographia* in that unbathed age. In the late twentieth century we have become blasé about the impact of scientific discovery, generally communicated by means of visual images, and it is hard for us to imagine the fascination value of a book like *Micrographia*, which opened up a hitherto invisible universe to the reading public.

One of the hallmarks of an outstanding scientific discovery in our own

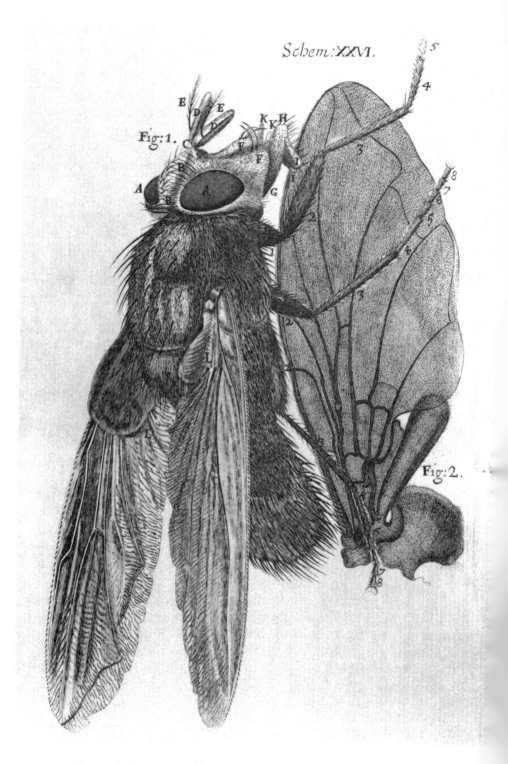

Fig. 2 The Blue Flye and its Wing from *Micrographia*

time—from black holes to DNA—is its influence on popular entertainment. *Micrographia* influenced the creative imagination of the Restoration in a variety of ways, but nowhere more embarrassingly, for Hooke, than in Thomas Shadwell's box-office success *The Virtuoso*, 1676. This play used the recently published discoveries and activities of the Royal Society to provide part of the plot motive and a main ingredient of the comedy in this farce of duplicity, seduction, and experimental philosophy. The butt of most of the jokes was Sir Nicholas Gimcrack, a foolish amateur scientist, or *Virtuoso*, who wasted his energies and fortune on seemingly absurd enterprises. Sir Nicholas 'spent two thousand pounds on microscopes to find out the nature of eels in vinegar, mites in cheese, and the blue of plums which he has subtly found out to be living creatures'. On 25 May, Hooke was told about the *Virtuoso*, then recorded in his Diary, on 2 June 1676, 'With Godfrey and Tompion at Play. Met Oliver there. Damned Doggs. People almost pointed.' [25] Such was the price of scientific fame.

Micrographia, therefore, was far more than a collection of careful observations made, as the title said, 'by the aid of magnifying glasses'. It was one of the first fruits of the new science to strike deep into the non-scientific imagination, and show how a cardboard tube containing two lenses could produce images of a vast new realm of knowledge. And when this realm was communicated through the medium of clear English and beautiful engravings, it could keep senior civil servants from their beds, and provide material for popular plays. Second only to Newton's *Principia*, which was a very different type of book that was published twenty-two years later, *Micrographia* was one of the formative books of the age, and assured Robert Hooke's reputation as a scientist of genius.

3. Of flight and of spring

Artificial flight, by means of mechanical contrivances, formed one of Hooke's most enduring fascinations, and could well have gone back to those childhood days in Freshwater when he devised clocks, model ships and other machines. 'At schoole he was very mechanical, and (amongst other things) invented thirty several ways of flying', [26] which must have amazed Dr Busby. And as in several other ways, he found an ideal *métier* at Oxford, for Dr Wilkins, the Warden of Wadham, had already published books on the use of flying machines, for both terrestrial and celestial journeys. With Dr Wilkins, Hooke 'made a module, which, by the help of springs and wings, rais'd and sustained itself in the Air'. [27] This led Hooke to consider a topic to which he would return at other times in his subsequent career: elasticity and spring.

After abandoning hope of using human muscle power in the propulsion of a flying machine, Hooke attempted to devise an artificial muscle substance, though we do not know what he used. As no rubber-based elastic materials were known in the seventeenth century, it is likely that his principal experimental agents were metal springs and gunpowder, which sometimes figure in the seventeen references to flight that occur in his Diary between 1673 and 1679. [28] But one Diary entry for 11 February 1675 not only makes a distinctly boastful claim for the invention of an artificial muscle, but indicates that his quest for such a material went back to his Oxford days:

> Dr Croon at Royal Society read of the muscles of birds for flying. I discoursed much of it. Declared that I had a way of making artificial muscule to command the strength of 20 men. Told my way of flying by vanes [wings] tryd at Wadham. [29]

Hooke's references to the use of gunpowder probably relate to the agent that was to be used to compress the artificial muscle prior to its activation. Various scientists around Europe were considering the force provided by controlled explosions as an agent that might perform mechanical tasks.

Robert Hooke left no clear drawings or even descriptions for either a complete flying machine or for any kind of 'artificial muscle'. His most detailed treatment of actual flight came from his descriptions of insects in *Micrographia*. In his account of the Blue Fly (Observation 38, p. 172) one finds Hooke's genius for observation, experiment, and mechanical analysis at its finest. What especially interested Hooke were the different wing-velocities of flies and bees on the one hand, and butterflies and moths on the other. Under the microscope, those insects that buzzed as they flew had hard 'glassy' wings, whereas moths were silent and had downy, or 'feathered' wings. These 'feathers', Hooke considered, could trap air and help the creature to float, so that less energy was needed to fly, and they made no sound.

The polished surfaces of bee and fly wings, however, were incapable of trapping air, so that only an exceedingly rapid motion could sustain them in flight, in consequence of which they buzzed. From these preliminary observations, Hooke went on to investigate the aerodynamics of the Blue Fly. Securing a lively fly upon the end of an unsharpened quill by means of a spot of glue, he first examined it with a powerful magnifying glass. When he tilted the quill, so that the fly might sense that it was flying in forward or vertical directions, he observed, from the blurred shape that they made, that the wings beat in arcs of different amplitude with relation to the direction of the fly's body. As the amplitude changed with assumed flying direction, so did the musical pitch of the buzz. Hooke suggested that a stringed instrument might be tuned in unison with this pitch, and that the ensuing note or vibration could be expressed in mathematical terms.

Very frustratingly, *Micrographia* takes this point of physical vibration and musical pitch (which was two centuries before Helmholtz) no further, other than to conclude that a fly's wings beat 'many hundreds if not some thousands of vibrations in a second minute of time' [30] and that it was the most rapid mechanical action in nature. But Hooke does appear to have come up with a relatively precise value within eighteen months of *Micrographia*'s publication, for Samuel Pepys, who was still very much of a scientific beginner, recorded in his Diary a meeting with Hooke on 8 August 1666:

> He did make me understand the nature of Musicall sounds made by strings, mighty prettily; and told me that having come to a certain Number of Vibracions proper to make any tone, he was able to tell how many strokes a fly makes with her wings (those flies that hum in their flying) by the note that it answers to in Musique during their flying.

Though Hooke does not explicitly state how he believed the wing-action of a fly kept it airborne, he seems to have related it to the spring, or elasticity, of the air, as the down-beat of the wings created a somewhat compressed pocket of air on which the creature rode, while the up-beat created a form of suction that lifted it up. One can see how these concepts related to his earlier work, conducted when he was Boyle's assistant, into the 'spring' of air.

It has already been mentioned that vibration was widely used as an explanatory mechanism by Hooke. Vibration was intimately related, in Hooke's thought, with elasticity, spring, and resonance, while these in turn were seen as mechanical responses to more fundamental agencies such as 'force', light, weight, and gravity.

While Hooke possessed no coherent idea of how springy bodies differed in structure from inert ones, it is clear that his scientific interest in elasticity went back to his Oxford days. It was in the company of Wilkins and Boyle that he had first encountered ideas of making artificial muscles for flying machines, and he recalled a curious pneumatic fountain that Dr Wilkins had built in the gardens at Wadham College, whereby

> the Spring of the included Air to throw up to a great height a large and lasting stream of water: which water was first forced

> into the Leaden Cistern thereof by two force pumps which did
> alternately work, and so condence the Air included in a small
> Room. [31]

As this device had been operational before Hooke had built the airpump
for Boyle, one wonders at the range of experiments with compressed air
that were going on in the 1650s, before it was possible to work properly
with vacuums.

Like many of Hooke's researches, his work on spring and elasticity
was done at scattered intervals stretching over several decades. In
the late 1650s, he had been experimenting in Oxford with spring-
regulated timepieces, and his 'pendulum watches' attempted to
apply the isochronal principle of the pendulum clock to a portable
timepiece in 1660. [32] His fullest discussion of the use of springs to
produce isochronal swings within a watch balance was published in
1676, and was intended to develop a timepiece whereby a ship could
find its longitude at sea. Though Hooke's timepieces were not sufficiently accurate to be used as marine chronometers (nor would such a
device be practicable for nearly a century), his application of spring tension and release to produce equal rotations of a watch balance-wheel
provided the fundamental principle on which all portable time-keepers
would be based down to the invention of electronic chip watches in our
own time.

It was in his *Helioscopes* in 1676 that Hooke followed the popular
seventeenth-century conceit of announcing a discovery in an anagram:
cediinnoopssttuu. He published its key two years later, in his most complete treatment of elasticity, in *De Potentia Restitutiva*, or *Of Spring*.
Here Hooke enunciated the original formulation of the law that bears his
name: *Ut Pondus sic Tensio*, or 'the weight is equal to the tension'. [33]
As the tension was seen as the product of an increasing series of weights
in pans suspended on coiled springs, it is easy in this pre-Newtonian-
gravitation age to understand how Hooke spoke of the *pondus*, or weight,
as acting on the spring. The formulation of 'Hooke's Law' with which we
are more familiar today is *Ut Tensio, sic Vis*, or 'the tension is equal to
the force'.

In *De Potentia Restitutiva*, Hooke presented his most complete
treatment of vibration and elasticity, as well as expressing his concept of
an aether that pervaded 'the whole Universe' and in which particles
moved continually. This concern with a vibrative agency that could
express the motions of a fly's wings and also the propagation of light, or
gravity, in space, was the nearest that Hooke came to a 'Grand Unified
Theory' in the Mechanical Philosophy. And considering his fundamental

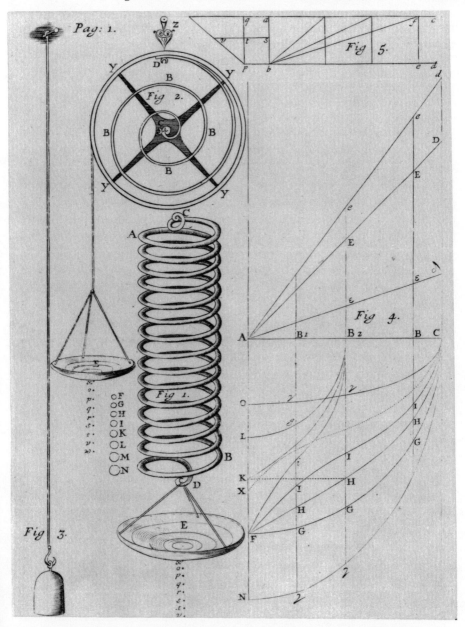

Fig. 3 Hooke's Spiral Springs, from *De Potentia Restitutiva*, 1678.

concern with *mechanism* in nature, one can understand the resentment that Hooke felt when Newton presented the very different unifying theory of Universal Gravitation to the Royal Society in 1686.

Before considering Hooke's ideas on gravity, however, it is important to look at his wider work in astronomy, and the way in which instrumentation and mechanism lay at the heart of his thought processes in this as in all the other branches of science that he investigated.

4. Astronomy and gravitation

Of all the individual branches of science to which Robert Hooke made significant contributions, astronomy was the most extensive. Astronomical matters concerned him, in one way or another, for well over forty years, as he dealt with the subject theoretically, observationally, and from the viewpoint of instrumentation. Of the twenty-one papers that he contributed to *Philosophical Transactions*, over a dozen deal with astronomy. Two astronomical observations appear in *Micrographia*, and during his frenetic decade of architectural activity, the 1670s, he produced his most significant astronomical publications, while his last recorded scientific investigation, in 1702, was an attempt to measure the solar diameter more accurately.

Hooke was an assiduous collector of data relating to the natural world, and he took particular pleasure in using the telescope and the microscope to add to it. The surfaces of planetary bodies were of great interest to Hooke, especially during the 1660s, when he was using various long telescopes (including those of twelve and sixty feet in focal length) to observe them. Robert Hooke, along with Cassini and Huygens, was among the first astronomers to carefully observe the surface of Jupiter. In May 1664, he reported a small round spot on the biggest Jovian belt, which he believed, unlike a satellite shadow, to be a permanent feature. It moved over two hours, and while Hooke later claimed to have used it to measure the planet's period of axial rotation, it was Cassini, in fact, who first published a value for Jupiter's rotation. In June 1666, however, Hooke reported another permanent spot, and differentiated its appearance from that of a satellite shadow. [34]

Observation 60 (p. 242) in *Micrographia* consists of an examination of a group of lunar craters made with a thirty-foot telescope in October 1664. In addition to the very considerable amount of detail that Hooke includes in his survey of that part of the lunar surface which Riccoli had named Hipparchus, he follows this with a discourse on lunar geology. As in most things, Robert Hooke is not satisfied with simply describing something, but is impelled to construct a plausible experiment and find an explanation. From his observations of the concave shadows produced inside the 'pits' or craters, Hooke proposed that they were probably the products of Earthquake (or 'Moonquake') generating processes within the body of the satellite. He likened the 'pits' to those found on the surface

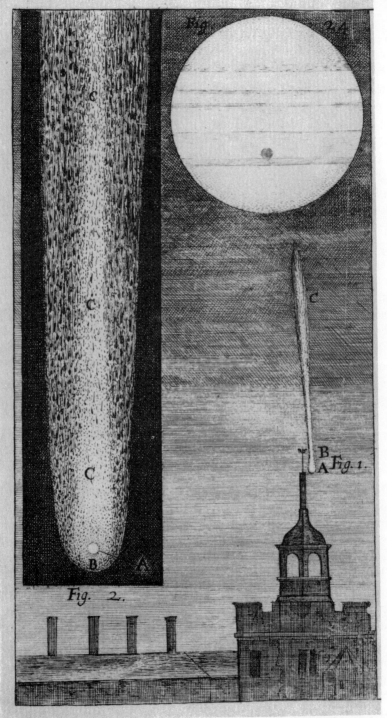

Fig. 4 The Comet, April 1677, from *Cometa, or Remarks about Comets* (London, 1678).

of a pot of boiling alabaster, or the outlines of the bubbles that remained if one blew air through a nearly solidified mixture of pipe-clay and water (p. 243).

Hooke's astronomical observations contain references to the characteristics of the telescopes with which he made them, as one might expect from a man who possessed such a thorough-going instrumental and sensory approach to research. But it is in his observations of the Pleiades star cluster, also recorded in *Micrographia* (Observation 59, p. 241), that he initiated discussion into what modern astronomers refer to as the 'resolving power' of the telescope. Though Hooke was aware of the higher magnifications obtained with object-glasses of longer focal length, and that with a telescope of twelve feet he could see seventy-eight stars in the Pleiades, whereas Galileo had only been able to see thirty-six, he recognized the crucial principle that it was object-glass diameter that was of primary importance in seeing faint objects. He experimented with a series of object-glass stops, and noticed that he saw the maximum number of stars through an entirely unstopped lens, though, surprisingly, Hooke never gives his object-glass apertures in inches.

During the winter of 1664–5, the skies of the northern hemisphere were dominated by a brilliant comet, which was the most conspicuous since that of 1618. Hooke observed it with his long telescope from London, and conducted the first detailed investigation into cometary nuclei and tails. It was not until a second brilliant comet appeared in 1677, however, that he published his researches. From studies of the two comets, Hooke concluded that their nuclei were solid bodies. A combination of their movement through the aether, and internal agitations within, caused the nuclei to be gradually eroded away, to form tails and streamers across space. He also concluded that comets did not merely reflect sunlight, but generated some light of their own, for there was never any sign of shadow in a cometary nucleus, even in those parts that were not facing the sun. [35]

The cause of cometary attrition and light-generation puzzled Hooke, and while he saw the aether as playing a role in this process, it was not so straightforward as that played by aerial nitre in the burning of a candle flame. In a cometary nucleus, observed Hooke, the brightest part was at the very centre, whereas in a candle or lamp flame (and here he drew attention to his work in *Lampas*), the centre was always dark, and the greatest light around it. [36]

When Hooke made his first observations of the comet of 1664, he lacked an effective micrometer whereby he could measure the angular diameter of the nucleus. His genius for improvization was displayed when he watched the comet low in the sky, and compared its nucleus diameter with the apparent width of an ornamental iron rod supporting a weather-cock on the roof of a distant building behind which the comet passed. [37] By later measuring the iron rod, and its distance from his telescope, he was able to calculate that the comet's nucleus was about 25 arc seconds, and its coma 4½ minutes in diameter.

How to obtain accurate angular measurements ran through almost everything Hooke did, for astronomy was the most intensely instrumental of the sciences, and astronomers across Europe realized that it was on refined angles that arguments and theories must stand or fall. Hooke's lack of a micrometer was solved in 1667, when he saw Richard Towneley's instrument, which was based on a prototype of 1640 invented by the Yorkshireman William Gascoigne. This instrument used a pair of fine-pitched screws to move two pointers in the focal plane of a Keplerian telescope. By enclosing the object to be measured between the pointers, its angular diameter could now be computed to within a few arc seconds, if one knew the exact focal length of the telescope, and the pitch of the screw which moved the pointers. Hooke published an engraving of the instrument to accompany Towneley's description in 1667. Its principle was to lie at the heart of astronomical measurement down to the twentieth century. [38]

Hooke quickly developed the concept of using screw turns to measure angles in the remarkable quadrant that he described to the Royal Society in 1674. In this instrument, he attempted to avoid the problems of unequally drawn degrees on the scale of an astronomical instrument by cutting fine teeth into the brass edge of the quadrant. By rotating a precision tangent screw along these teeth, he hoped to be able to express degrees, minutes, and seconds in full and part turns of the screw. It was a brilliant and portentous idea and one of the earliest attempts to apply precision mechanics to astronomy. Unfortunately, like so many of Hooke's inventions, it went beyond the current skills of manufacture and failed to work properly. But one Hooke invention that was originally intended to form part of an astronomical instrument was his celebrated 'Universal Joint', which was devised to operate an adjusting arm of his *Helioscope* apparatus in 1676. [39]

One man who tried to use a Hooke screw-edge angle-measuring instrument for regular astronomical research was the first Astronomer Royal, John Flamsteed, for Hooke had been brought in to provide designs for some of the original instruments of the Greenwich Royal Observatory in 1675. But it is clear from his remarks that Flamsteed did not find Hooke's prototype instrument to be very successful in practice, complaining that he was 'much troubled with Mr. Hooke who, not being troubled with the use of any instrument, will needs force his ill-con-

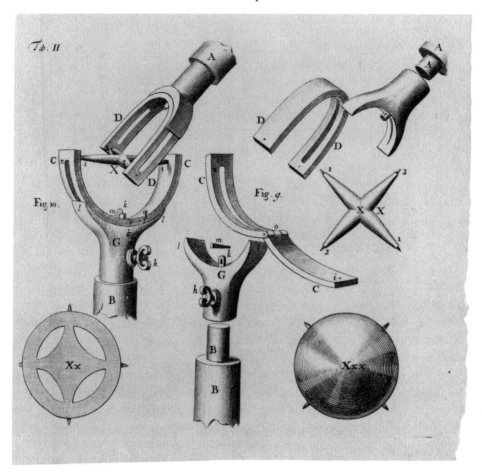

Fig. 5 Hooke's Universal Joints, from *A description of Helioscopes*, 1676

trived devices upon us'. [40] Robert Hooke was the brilliant deviser of machines, who could detect a way whereby a principle in mathematics could be expressed in metal to produce an experimental instrument. Flamsteed, on the other hand, was the painstaking working astronomer, going through his nightly slog of measuring star positions on top of Greenwich Hill, and he did not wish to be burdened with experimental instruments, no matter how cleverly they were conceived.

But John Flamsteed owed more debts to Robert Hooke than he cared to acknowledge. He made extensive use of telescope micrometers at Greenwich, while Hooke's devising of a thirty-six-foot zenith instrument in 1669, in his attempt to measure a stellar parallax, provided an interesting application of telescopic mechanics to trying to prove the Copernican theory. Hooke's endeavour to demonstrate the motion of the earth failed, but this only led Flamsteed to try for himself, and also fail; and he further took up Hooke's suggestion that zenith star images could best be observed from the bottom of deep shafts in the ground, setting up his 'well telescope' at Greenwich. [41]

Gravitation was a subject which occupied Hooke's interests for well over twenty-five years before Newton published his *Principia*. According to Aristotle, the Earth could be the only gravitational body, as it drew 'heavy' things towards itself. By the 1660s, however, astronomers were considering the possibility that gravity could be possessed by other bodies as well, such as the planets. The nature of gravitation was in itself mysterious, though in his lunar observations in *Micrographia* (p. 245) Hooke had argued that the even, spherical shape of the moon indicated that it must possess a gravitational power which caused everything to fall evenly around its centre. The same went for the planets.

As a mechanist, Hooke looked for physical connections in nature through which gravity might operate, and this led him to believe that an aether must exist whereby it (along with light and magnetism) could move or resonate. His thorough-going experimentalism was always leading him to new 'tryalls', and in 1662 and 1665 he reported to the Royal Society experiments conducted on the towers of Westminster Abbey and old St Paul's Cathedral. Identically heavy weights on identically heavy lines were prepared. One was rolled up and placed in one pan of a balance, while the other was set free down the tower while attached to the other balance arm. Would the weight that was now 90 feet closer to the earth become heavier against its rolled-up companion? He found no appreciable change, so he tried it in reverse, as it were, down a deep well on Banstead Downs, in Surrey. Once again, no decisive results could be obtained. [42]

By the 1670s, he was trying to find the operation of gravity in space. He argued that the motions of comets, whereby they were pulled out of once stable orbits to move near the sun and earth, must be occasioned by the 'vortices' through which gravity operated. But it was in his *Attempt to prove the Motion of the Earth* in 1674 that Hooke made some of the most pertinent remarks about gravitation that were made before Newton. He summed them up under three headings: (1) that gravitation exists towards the centres of all bodies, and between all bodies; (2) that all bodies will move in straight lines under their own impulse, but can be disturbed into orbits by other gravitational bodies; (3) that gravity acts more powerfully when bodies are closer together than when further apart, but 'what these several degrees are I have not yet experimentally verified'. [43]

He was clearly much closer to a solution by 6 January 1680, when he told Newton that gravitational attraction 'is always in a duplicate proportion to the Distance from the Center Reciprocall', and 'as Kepler Supposes Reciprocall to the Distance'. Hooke does not tell us exactly how he had come to these all-important conclusions, though it is likely that it was by a combination of astronomical observations and the experiments with iron balls which he mentions in the same letter. He also told Newton in the same letter that Edmond Halley, who had recently returned from St Helena, had been puzzled by the fact that his pendulum clock ran slower up the mountain than it did when lower down, 'But I presently told him [Halley] that he had solved me a query I had long Desired to be answered ... that ... gravity did actually Decrease at a height from the Center.' [44]

It was Newton's refusal to acknowledge Hooke's insight into this Inverse Square Law of Gravitation, following the publication of *Principia* in 1687, that led to the appalling débâcle which broke out in the Royal Society. Yet with a knowledge of Newton's work to hand, one can see how very differently the two men approached the problem of gravitation, and science in general. Hooke was the mechanist, constantly searching for physical connecting agents that could be demonstrated experimentally. Newton was the mathematician, willing to accept the presence of force that acted at a distance, provided that it was amenable to precise mathematical expression.

If astronomy, in its various branches, displayed Hooke's creative genius and powers of investigation at their highest, it was the practice of architecture that made him a rich man.

Robert Hooke, architect and City surveyor

It may seem strange to us today that a man without any practical training in building or architecture should have been appointed by the City of London authorities to be their Surveyor following the Great Fire in 1666. We know that Hooke was quick off the mark in presenting a proposed new ground-plan for the City, almost as soon as the embers were cold, but we do not know who pleaded his case for the Surveyorship. His old Oxford encourager, Dr John Wilkins, could well have been one of his backers, for Wilkins was well connected in the City, but we have no certain knowledge.

But the appointment of a non-professional like Hooke would not have seemed so absurd as it would do today. There was, after all, a tradition of 'gentleman' architects three centuries ago, with figures like Sir Roger Pratt, and especially Hooke's friend, Christopher Wren, who was appointed to the parallel post of Surveyor to the King. Between them, these two scientists, Hooke and Wren, were to be responsible for designing most of the principal Royal and City buildings in the metropolis. In Wren's case, the commitment of time was to shift away from mathematics and science to architecture almost full-time by 1675, though Hooke ran his Surveyorship and architectural practice in conjunction with an unabated commitment to scientific research. [45]

What made the untrained 'gentleman' architect a serious figure in the seventeenth century was the nature of the academic education that he would have received. Classical architecture, in many ways, is very formulaic, with its Orders, Cubes and proportions. A man with a knowledge of Latin and Euclidean geometry could master Vitruvius' *De Architectura Libri Decem* along with the sixteenth-century Italian architectural writers like Palladio. If he had a good eye for proportion, and a natural good taste that he had perfected by the close study of engravings of Greek, Roman and Renaissance buildings, he could learn to produce elegant designs. And if he had the imaginative powers of Wren and Hooke, combined with their scientific grasp of thrusts and forces, then he could produce architecture of genius. While Hooke's architecture may not have been quite so brilliant as Wren's (and far fewer examples of it survive), engravings of Physicians' College, Bedlam Hospital, and other edifices none the less show that he was genuinely gifted as an architect. These gifts also found concrete expression in the very substantial fortune that he made from architecture and City planning.

Robert Hooke's Diary in the 1670s gives one some indication of his commitment to the Surveyorship. He had to authorize the safety aspects and street widths of other mens' designs, to try to make the new City less fire-hazardous than the old one, while he accepted commissions to provide drawings of his own. The rebuilding of London must also have put Hooke on his mettle as an administrator, as he tried to implement safety regulations, design buildings that pleased fee-paying clients, and organize tradesmen and contractors to see that things took physical shape. [46]

In London, Hooke designed many Livery Company Halls, part of the Thames and Fleet waterfronts, private residences in St James' Square, and great mansions, like that for Lord Montague. Outside London, he designed Ragley Hall, Warwickshire, almshouses, and parts of the Tangier Mole: numerous buildings of virtually every architectural rank. He sometimes worked in conjunction with his friend Wren, most notably on the design of the Royal Observatory, Greenwich, and the Fire Monument, or 'Piller', in Fish Hill Street. But his greatest independent creations were the magnificent buildings for the Royal College of Physicians, completed in 1679, and Bethlehem Hospital, or 'Bedlam'. [47]

Physicians' College, with its 'gilded pill'-surmounted dome, later came to be popularly ascribed to Wren, until Hooke's long-lost Diary came to light in Guildhall Library around 1890. Bedlam Hospital, with its French château style, pavilions, and 540-foot façade, was magnificent by any standards. The wits of the day made the remark that the English housed their lunatics in buildings such as those in which the French housed their Kings! In the original design, Bedlam represented advanced thinking in its accommodation of the mentally ill, with an individual cell for each inmate, which led the comic writer Ned Ward to comment upon the hospital's governors 'I think they were mad that built so costly a College for such a crack-brained Society'. [48] In Bedlam, Hooke had not only produced his own architectural masterpiece, but had created one of the landmarks of London for the next century. It figures in Pope's *Dunciad*, foreign visitors went to see the antics of the unfortunate patients, and William Hogarth set his last scene of the *Rake's Progress* in its now squalid and over-crowded interior. It was demolished in the early nineteenth century, and all that now survives are the two statues that once flanked the entrance: the frightening images of the 'crazed brothers', *Raving Mania* and *Melancholy Mania*, who now reside in the Victoria and Albert Museum.

Robert Hooke's buildings have had a disastrous survival record in general, which is probably one reason why we generally do not speak of him in the same context as Sir Christopher Wren. Most of what had survived into the nineteenth century perished in the wholesale remodelling of Victorian London. Even the Second World War took its toll, when the magnificent wooden screen which he designed for the Company of Merchant Taylors was destroyed by a bomb. The only Hooke building that survives intact is the beautiful red-brick parish church at Willen, Buckinghamshire, which he designed in 1680 for his old Head Master, Dr Busby, who was patron of the living. It is well worth a visit, in greater Milton Keynes.

In the 1670s Hooke showed himself to be a skilful architect, capable of working in several styles. He was also a competent administrator and planner, and played a major part in raising London from the ashes of 1666. And all of this was going on at the same time as he was measuring the stellar parallax, experimenting on respiration, flames, and vibrations, improving the design of watches, and formulating *Ut Pondus, sic Tensio*.

Hooke's character and friends

Robert Hooke never married, and on being appointed to his Gresham Professorship in 1665 was happy to live in College to the end of his days.

Though his first biographer, Waller, presented him as a melancholic, 'monastick' old bachelor, one must remember that this was the elderly, embittered Hooke who felt cheated by Newton. For most of his life, as his Diary vividly exemplifies, he was a highly clubbable man with a large social circle, though the mainsprings of his life, and of his friendships, were intellectual.

Though he never married, Hooke's Diary makes it clear that he had fleeting liaisons with a succession of seamstresses and maidservants. [49] But he was generous to them, and occasionally helped them to find husbands. Indeed, generosity seems to have been a clear mark of the younger Hooke, combined with a genuine kindness towards those who were dependent on him. These include the orphan boy, Tom Gyles, whom Hooke took in and educated, the young Harry Hunt, who later became Curator of Experiments at the Royal Society, and his ward, Grace Hooke. His friendships, moreover, seem to have covered a large social spectrum, and included enterprising tradesmen as well as Fellows of the Royal Society. Hooke's endlessly hungry mind crossed all social barriers in search of nourishment.

John Aubrey was a long-standing friend. Like Hooke, he had a roving intellectual appetite, but unlike him, was a child in the ways of the world. The Wiltshire landed gentleman, sliding inexorably into bankruptcy, contrasted sharply with the curate's son who was also an astute man of business, but they got on excellently. Sir Christopher Wren was also deeply valued and admired by Hooke, as was Boyle. Robert Hooke was a man of passionate likes and dislikes and he never forgot a friend or benefactor. He lamented the death of Dr Wilkins in 1672, and stayed in contact with Dr Busby down to his old Head Master's death in 1695.

His highly strung, prickly pride made it very natural for him to snap and scratch if he felt slighted or patronized. He disliked Henry Oldenburg ('Kindle-Cole', or mischief-maker) in the Royal Society, while he and Flamsteed were like cat and dog with each other. His traumatic relationship with Newton after 1687 seemed to overshadow the rest of his life, making him 'Melancholy, Mistrustful, and Jealous, which were increas'd upon him with his years'. [50] This was most unfortunate, for by temperament Hooke was an open and direct man.

During the last couple of years of his life, Hooke's health deteriorated rapidly. He suffered from swollen legs, chest pains, dizziness, insomnia, extreme emaciation, blindness, and what was possibly gangrene of the feet. Though it is impossible to diagnose causes of death across nearly three centuries, one suspects the presence of cardiovascular disorders and possibly diabetes as contributory factors. Hooke made no will, though he left £9580 in money and goods, along with some small property on the Isle of Wight. It was an estate that many country squires would have been proud to leave. He was buried in St Helen's Church, in the City, with all resident Fellows of the Royal Society in the cortège [51].

Robert Hooke was a figure of extraordinary and diverse creativity. With his grasp of ancient languages, the quality of his draughtsmanship as shown in the plates of *Micrographia*, and his success as an architect, he clearly possessed high artistic talents. And his craft skills enabled him to build an airpump where the country's leading pumping engineer had failed. But most of all, he was the man who showed that the 'experimental philosophy' actually worked and could be used to extend the bounds of natural knowledge. He was Europe's last Renaissance man, and England's Leonardo.

Acknowledgements

I wish to thank Mr A. V. Simcock, of the Museum of the History of Science, Oxford, for his assistance with books and early printed sources. I also thank Mrs Sheila Edwards and her staff at the Royal Society Library, and Mrs Irene McCabe and Dr Bryson Gore of The Royal Institution for their assistance with the Discourse.

Notes and References

1. Z. C. von Uffenbach, *Merkwürdige Reisen durch* . . . Ulm, 1753. Translated as *London in 1710* (trans. W. H. Quarrell and M. Mare), London, 1710, p. 102. *Aubrey's Brief Lives* (ed. O. L. Dick), London, 1975, p. 165. *The Posthumous Works of Robert Hooke* (ed. R. Waller), London, 1705, p. xxvi.
2. Robert Hooke, *Micrographia, or Some Physiological Descriptions of Minute Bodies Made by Magnifying Glasses with Observations and Inquiries thereupon*, London, 1665, 'Preface', unpaginated, pp. 7–8 from beginning.
3. Hideto Nakajima, 'Robert Hooke's family and his youth: some new evidence from the will of the Rev. John Hooke', *Notes and Records of the Royal Society*, **48** (1); (1994), 11–16.
4. Most of Hooke's early biographical information comes from Aubrey's and Waller's accounts (see ref. 1). See also R. E. W. Maddison, *The Life of the Honourable Robert Boyle*, London, 1969, pp. 92ff.
5. Thomas Sprat, *A History of the Royal-Society of London*, 1667, 2nd edn, London, 1702, p. 67.
6. Margaret 'Espinasse, *Robert Hooke*, London, 1956, p. 83.
7. Robert G. Frank Jr., *Harvey and the Oxford Physiologists*, Berkeley, CA, 1980, pp. 133, 135.

8. Hooke discusses his 'aether' in a variety of places; e.g. *Posthumous Works* (ref. 1), pp. 172, 174, 184. For a modern study, see John Henry, Robert Hooke, incongruous mechanist, in *Robert Hooke, New Studies* (ed. M. Hunter and S. Schaffer), Woodbridge, 1989, pp. 157–62.

9. *Micrographia* (ref. 2), 'Preface', unpaginated, p. 3 from beginning. One of the most significant studies on Hooke and the role of instrumentation in his science is by J. A Bennett, Robert Hooke as mechanic and natural philosopher, *Notes and Records of the Royal Society*, **35** (1), (1980), 33–48.

10. *Oxford Physiologists* (ref. 7), p. 116. Charles Webster, The discovery of Boyle's Law, and the concept of the elasticity of air in the seventeenth century, *Archive of the History of Exact Sciences*, **2** (1965), 441–502. Dr Webster's article provides one of the most thorough studies of Boyle's pneumatic work. For his treatment of Roberval, see pp. 448–51.

11. Robert Boyle, *New Experiments Physico-Mechanical, Touching the Spring of the Air, and its Effects (Made, for the most part, with a New Pneumatical Engine)* . . ., 2nd edn, Oxford, 1662, p. 4. Waller, in *Posthumous Works* (ref. 1), iii, states that Greatorex's machine had been 'too gross to perform any great matter'.

12. *New Experiments* (ref. 11), p. 8.

13. *New Experiments* (ref. 11), pp. 5–6.

14. *New Experiments* (ref. 11), pp. 10–11, 40–1.

15. Thomas Birch, *History of the Royal Society*, Vol. II, London, 1756, p. 10, 25 January 1665. *New Experiments* (ref. 11), Experiments 4 and 6, p. 32.

16. R. Boyle, *Certain Physiological Essays* . . ., London, 1661, pp. 107–10.

17. J. R. Partington, *A History of Chemistry*, Vol. II, London, 1961, p. 308. Also R. J. Frank Jr., John Aubrey F.R.S., John Lydall, and science at Commonwealth Oxford, *Notes and Records of the Royal Society*, **27** (1973), 193–217; for *aurum fulminans*, see pp. 197–8.

18. *New Experiments* (ref. 11), pp. 48–50. When Boyle did succeed in focusing sunlight on gunpowder, however, he found that it did not combust as rapidly as it did in air, but that *aurum fulminans*, when dropped onto a heated plate inside the airpump receiver, dissolved in a flash: *History of Chemistry*, Vol. II (ref. 17), p. 527.

19. R. Hooke, *Lampas: or, Descriptions of some Mechanical Improvements of Lamps & Waterpoises. Together with some other Physical and Mechanical Discoveries*, London, 1677, pp. 4–8. Hooke had already described his work on flame in February 1671, in 'An Experiment to prove the substance of a Candle or Lamp is dissolved by the Air, and the greatest part thereof reduced into a Fluid, in the Forme of Air', 14 March 1671–2, Royal Society MS, RBC 3, 201–3.

20. Robert Boyle, *Philosophical Transactions*, **5** (1670), 2043. For John Mayow see *History of Chemistry*, Vol. II (ref. 17), p. 602.

21. From *The Ballad of Gresham College* (anon.), a Broadsheet of *c.* 1663, printed with critical notes by Dorothy Stimson, *Isis*, **18** (1932), 103–17.

22. *Robert Hooke* (ref. 6), p. 51. Hooke also reported the effects of air deprivation on fish, on 20 May 1663, Royal Society MS, RBC 2, 31–32, and on 24 October 1667 on a dog, *Philosophical Transactions*, **2** (28), (1667), 539–40.

23. This was in John Mayow's *Tractatus Quinque*, Oxford, 1674; see *History of Chemistry*, Vol. II (ref. 17), pp. 587–613.

24. *The Shorter Pepys* (ed. Robert Latham), London, 1986, 21 January 1665, p. 464.

25. Thomas Shadwell, *The Virtuoso*, 1676 (ed. Marjorie Hope Nicholson and David Stuart Rhodes), London, 1966; Act 1, Scene II, line 6, p. 22. (In addition to microscopy, *The Virtuoso* made fun of 'the art of flying', blood-transfusion, and other Hooke interests.) *The Diary of Robert Hooke, M.A., M.D., F.R.S., 1672–1680* (ed. H. Robinson and W. Adams), London, 1935, 25 May and 2 June 1676.

26. *Brief Lives* (ref. 1), p. 165.

27. John Wilkins, *Mathematical Magick*, London, 1648. *Posthumous Works* (ref. 1), p. iv.

28. The best study of Hooke's work as represented by the *Diary* is in Richard Nichols, *The Diaries of Robert Hooke, The Leonardo of London, 1635–1703*, Lewes, 1994, pp. 169–72.

29. *The Diary of Robert Hooke* (ref. 25), 11 February 1675.

30. *Micrographia* (ref. 2), p. 173. Hooke also looked at the relation between vibration and music in 'A Curious Dissertation concerning the Causes of the Power & Effects of Musick . . .', an undated paper posthumously communicated by Dr W. Derham, 14 December 1727, Royal Society MS RBC 13.3. In this paper, Hooke argued that an awareness of the vibrations of music precedes an awareness of language, as babies respond to music (p. 6).

31. R. Hooke, *De Potentia Restitutiva, or of Spring Explaining the Power of Springing Bodies*, London, 1678, p. 23.

32. *De Potentia Restitutiva* (ref. 31), p. 5. See also Michael Wright, Robert Hooke's Longitude Timekeeper, in *Robert Hooke, New Studies* (ref. 8), pp. 63–118.

33. Robert Hooke, *A Description of Helioscopes and some other Instruments*, London, 1676, p. 32, item 9. Also *De Potentia Restitutiva* (ref. 31), p. 5.

34. See *Philosophical Transactions*, **1** (1), (1665–6), 3; *Philosophical Transactions*, **1** (14), (1665–6), 239–42 and 245–6 (mis-numbered 145). Also R. Hooke, *Cometa, or Remarks about Comets*, London, 1678, Supplement 'The Period of Revolution of Jupiter upon its Axis . . .', pp. 78–80.

35. *Cometa* (ref. 34), pp. 9–10, 32–4.

36. *Cometa* (ref. 34), p. 47. Also 'Of Comets and Gravity', in *Posthumous Works* (ref. 1), pp. 166ff.

37. *Cometa* (ref. 34), pp. 4–5.

38. Richard Towneley, A Description of an Instrument for dividing a Foot into many thousand parts, and thereby measuring the diameters of planets to great exactness, *Philosophical Transactions*, **2** (1667), 541–4.

39. R. Hooke, *Some Animadversions on the First Part of Hevelius, his 'Machina Coelestis'*, London, 1674, p. 55. *Helioscopes* (ref. 33), plate 11, and p. 18.

40. Letter, J. Flamsteed to R. Towneley, 3/7/1675, Royal Society MS, 243 Fl. 8.

41. *The 'Preface' to John Flamsteed's 'Historia Coelestis Britannica 1725'*, edited and introduced by Allan Chapman, based on a translation by Alison Dione Johnson, National Maritime Museum Monograph No. 52, 1982, pp. 179–80. For Flamsteed's 'Well Telescope' see Derek Howse, *Francis Place and the Early History of The Greenwich Observatory*, New York, 1975, p. 58, plate XII.

42. R. Hooke, 'Of the Difference of Gravity by removing the body further from the Surface of the Earth', 24 December 1662, Royal Society MS, RBC 1.

288–291. Also R. Hooke, 'Of Gravity', 21 March 1665–6, Royal Society MS, RBC 2. 223.

43. R. Hooke, *Attempt to prove the Motion of the Earth*, London, 1674, p. 23. The Essay 'Of Comets and Gravity' in *Posthumous Works* (ref. 1) provides a vibrative model of gravity, pp. 184–5.

44. R. Hooke to Isaac Newton, 6 January 1679–80, in *The Correspondence of Isaac Newton*, Vol. II, Cambridge University Press, 1960, p. 309. In his *Brief Lives* (ref. 1), Aubrey clearly stated Hooke's priority in recognizing the Inverse Square Law of Gravity, p. 166.

45. *Robert Hooke* (ref. 6), pp. 83–105.

46. *The Diary of Robert Hooke* (ref. 25) contains numerous references to architectural work during the 1670s. See also *The Diaries of Robert Hooke* (ref. 28), pp. 101–10.

47. For the most detailed study of Hooke's architecture, see Marjorie Isabel Batten, The architecture of Dr Robert Hooke, F.R.S., *Walpole Society (London)*, **25** (1936–7), 83–113.

48. Ned Ward, *The London Spy*, London, 1703, Folio Society Edition, London, 1955, p. 48.

49. Lawrence Stone, *The Family, Sex, and Marriage in England, 1500–1800*, London, 1977, pp. 561–3.

50. *Posthumous Works* (ref. 1), p. xxvii.

51. 'Hooke's possessions at his death: a hitherto unknown inventory', in *Robert Hooke, New Studies* (ref. 8), p. 294.

ALLAN CHAPMAN

Born 1946 in Manchester, obtained a first Degree from the University of Lancaster, then went to Wadham College, Oxford, for postgraduate work. A historian by training, working in the history of science, his particular research interests are in scientific biography and astronomy. He teaches the history of science in the Faculty of Modern History, Oxford. In addition to published research, he lectures extensively in the history of science in England and abroad and, in January 1994, gave the Royal Society's triennial Wilkins Lecture in the History of Science, on Edmond Halley as a historian of science.

Cells, functions, relationships in musical structure and performance

ROSALYN TURECK

Introduction

Music, the most abstract of the arts, embodies complexities far beyond its conventional definitions of form and structure. Because its notated symbols are translated into sounds and conveyed to our minds and sensibilities via a faculty of the senses, one is easily misled into regarding sounds *per se* as the primary characteristics of music. Whereas sounds, and silence, make up the raw materials of music, that which creates music occurs outside of sound. It is the organizing process of sounds— any sounds—that causes sounds to emerge as a form. Different organizing processes fashion sound and silence into various forms and structures. The result is what we term music.

My performance contains within a microcosm the sum total of my life's work and thought. It expresses a number of underlying concepts— a concept of form, of structure and of authenticity. But one does not ordinarily go to music performances to learn about concepts. This applies to even the most sophisticated of listeners. By concepts I do not mean the traditional formalized notions about musical form that provide the theory, rules and terms in Western musical composition, such as sonata or fugue form; nor do I include the prevailing idiomatic structural procedures of a culture such as those that characterize contrapuntal or harmonic musical organization. I refer to abstract styles of thought. These are more often associated and expressed with words and, in many cases, with mathematics. But form and structure in music eloquently reveal their conceptual components. If one can perceive the relationship of the structural language to the formal framework, the underlying concepts rise to the surface.

Although diverse aspects of this article are applicable to varied styles of music and to areas outside of music, the focus here relates to the structural *Weltanschauung* and specific organizing processes in the music of Johann Sebastian Bach. The fugue (BWV 866) in B flat major, from Book I of the *Well-tempered Clavier*, is the model for demonstrating the identities of the varied structural processes and relationships that are responsible for the formulation of a musical organism. The primary concern here is not the performance media or period performance practices, but rather the perception of Bach's structural processes, the abstract implications that these represent, and the potential for performance that these reveal. The fugue in B flat major from Book I, from which I extract examples for the charts that follow, is selected for analysis because it is extraordinarily condensed. Its materials are made up solely of three motives; no other material is introduced. The entire fugue in B flat major becomes an organic whole as the result of *positioning* the three motives in a framework of vertical simultaneity and in the multiple relationships caused thereby.

Musical symbols are actually more concise than verbal terms. These symbols depict distinct structural patterning and relationships that would require hundreds of words to identify and describe. I can present here only a tiny glimpse into the *structural* aspect. I cannot begin to touch on the concept that underlies the structure intrinsic to the form of the music and of my performances. But I hope that from this verbal exposition and the illustrative charts it can be perceived that musical structure contains a potential in its vast range of relationships and processes that has hardly been tapped. And this potential, awaiting analytic perception by the performer, forms one of the sources that nourish my performances.

Niels Bohr once said something to the effect that when one thinks one has mastered a subject, one then writes a book that is so obscure that nobody can understand it. However, I will attempt to project my theories clearly in the visual and verbal modes just as I try to fulfil the interior elements and relationships, as well as exterior aspects of the music, in my performances.

Before turning to specific points that emerge from my method of analysis or performance, I should like to quote a mathematician friend of mine, Professor Henk Barendregt of Nijmegen University in the Netherlands, describing a generalized notion of structure from a mathematician's view. [1]

> The notion of structure as used in mathematics stands for a collection of objects together with some given relations between these ... These objects can be anything and the relationship between them may be arbitrary; how are such structures meaningful and interesting?

> One way for a structure to be interesting is that it may satisfy some properties that are defined in terms of the structure itself (internal properties). A second way to be interesting is that a structure can be related to other structures, of which the interest has been established by other means (external properties).
>
> Classical mathematics is concerned with the study of a few fixed structures and their internal properties. Modern mathematics (starting around the early 19th century) is concerned with *classes* of structures satisfying some fixed internal properties and it studies external and other internal properties of these . . .

This statement can be paraphrased for application to structure in music. The 'internal properties' of musical structure are the areas of my detailed analysis. But the examination of internal properties leads to consideration of their relationships to each other and to the whole that *contains* them. The 'whole' may be anything from a two-note motive to the entire composition. Traversing this entire gamut is my way of analysis—from the smallest unit through to the entire form, and the relationships activated by the structure's materials and processes. Since so many factors operate simultaneously, I do not approach music analysis in the conventional linear fashion.

My work in structure emerged from my interest in penetrating Bach's music for my own comprehension as an artist and for purposes of performance. The results of my analyses of levels of structures and their relationships led me to delve more deeply into the diversity of structural processes in Bach's music. This then led me to thinking about varying forms and structural devices in the music of different eras and cultures. These probings drew me beyond conventional styles of music analysis. I then found it necessary to consider the style of *thought* that shaped the sense of form and dictated the *structural idiom* of the diverse musical forms that were produced in different eras.

This then, in brief, is the genealogy of my work in the abstract realm of music, which includes the conceptual foundation, the framework that the fundamental concept provides for specific forms and structures, the analysis and synthesis of these and, ultimately, their amalgamation into the interpretative and technical devices that are stylistically pertinent to practices and associations of specific musical cultures.

I analyse music according to its own behaviour and not according to contemporary analytical methodologies or performance approaches. These vary not only from era to era, but even from decade to decade. My performance applications are also based in the behaviour of musical structures. My perception of the behaviour of musical structures according to their conceptual and individually unique nature necessitated the creation of performing techniques corresponding to the material, the structural processes, the context of these structures and their processes, and their individual and overall relationships.

The conceptual foundation underlies the realm of performance as well as that of composition; moreover, multilevels of structure and their relationships are present also in structuring performance. For, a musical performance is like a composition, a structure in sound arising from a fundamental, abstract concept. Like musical composition it is an amalgam of varied elements and processes. In addition to the varieties of sonorities, textures, and their immeasurable nuances, which provide us with sheer aural joy, are the building blocks that make up *forms* of sounds and silences. The *process* of fashioning the building blocks of performance into structured relationships breathes life into the sounds and silences of *compositional* forms and structures. Compositional processes in music function as its organizing elements. Performance devices that emerge from musical forms and structures function as direct reflections of the organizing elements of the composition.

Form and structure are not the unique possessions of the composer; they are equally present and needful in performance. In actual fact, every performer creates some sort of a structure in sound, whether the performance is the product of naiveté or highly developed sophistication. *A musical performance constitutes a system moving in time; it takes shape in the silence of time*. Depending on the degree of awareness on the part of the performer, the form of a performance may vary from amorphous to densely structured. The choice and the structuring of performance devices, whether few or many, may or may not coalesce with the structuring of the composition that is being performed. If not, anachronisms and distortions of the original structures occur. My approach, identifying performance structuring with compositional structuring, has to do with the *composer's* sense of form and specific structuring, not with the instrument employed or on externally imposed temporal fashions, either of a past era or of the current one.

Performance devices are too frequently regarded as apart from compositional devices, being identified chiefly with the characteristics of instruments. Additionally, in our time performance devices are strongly linked with performing practices within a designated culture. There is no denying the magnetic pull of a culture's *Weltanschauung* as well as its technological products such as instruments and their respective sonorities and techniques. Yet Claude Debussy, who is universally associated with the characterization of evanescent mood and mysteries of instrumental tone painting, said [2]

> Music is the arithmetic of sounds as optics is the geometry of light.

This is the stark, demystified reality of creative art; a composer, even such a tone and mood painter as Debussy, defines music as 'arithmetic of sounds', rather than in terms of sonorities, textures, and volume of sound.

In the conscious, and conscientious, study of a musical composition, a clear delineation of characterization emerges as the *result* of structural analysis, not before. Mood and characterization, and decisions about tempo, although usually the first concern of the performer, reveal themselves gradually in all their strengths and nuances as the structures yield up, stage by stage, the diverse levels of their content and processes. This is not, as may be supposed, an arid, detached intellectual exercise; quite the contrary, intuitive perception also plays a part in this kind of investigation. Moreover, the area for emotional responses is greatly expanded as a consequence of an enlarged perception of multilevelled structures and processes.

I shall be referring to terms and materials that will be familiar to most of my readers—time values, barlines, time signatures, etc. But what I have to say about the musical formation of these materials in a composed piece of music will be unfamiliar, especially in relation to the coordination of these materials with performance.

My method of analysing structure in composition exposes structural treatments for performance. Above all, I perceive performance as constituting a structure in itself. And a performance structure can become extremely complex, as it identifies on every level with the structuring of a complex composition. Where a performance does not identify on multiple levels with the multiple levels of a composition, I hold that authenticity in performance is lacking. It is the music of the composer that should be the primary concern of the performer. Certainly this is the primary concern of the composer when evaluating a performance!

In using the term music, I refer to music that is composed. Composed music may be any group of organized sounds, from a primitive succession of sounds ordered in a way unique to a culture in order to fulfil a particular function or ritual, to the complex structures and form embodied in Johann Sebastian Bach's *Art of Fugue,* or Alban Berg's operas *Wozzeck* and *Lulu.* Although most music developed a written notation, even chants that are passed on orally from generation to generation are composed using specific, established patterns (generally classified as traditional) and patterned relationships.

1. Symbols

Composers communicate via music notation, which consists of symbols unique to music. These symbols vary with different musical cultures.

Musical literacy consists, in its initial stages, of recognizing the alphabet of symbols and their grammatical functioning within the musical form.

Notation

Two modes of notation—Variation 13, J. S. Bach, Goldberg Variations.

Illustration 1 shows a notation style that reigned for several centuries previous to the mid- to late-eighteenth century. This notation is not

Illustrations 1 and 2 Two modes of notation for Variation 13 of J. S. Bach's Goldberg Variations.

Variatio 13

ILLUSTRATION 1

ILLUSTRATION 2

Variatio 13

written in the conventional large notes on the staff. Its symbols represent chiefly figures—groups of notes rather than single notes, although there are signs that represent single notes as well when required. I have reduced the figuration of the variation as shown in conventional modern notation (see Ill. 2) to the earlier mode of notation symbols which were conventionally employed for what is termed 'embellishment'. The conventional European notation symbols of roughly the last three hundred and fifty years are shown in Illustration 2, on the alternating staves of the treble and bass clefs. The embellishment symbols represent the figurations notated via the large notes on the staff in the treble clef. I have placed the reduction of the figuration into embellishment symbols (Ill. 1) directly above the realized figurations of these symbols in order that the visual correlation of embellishment symbol notation and notetext staff notation may be easily perceived.

Embellishment notation, being capable of incorporating a figuration of two or more notes within one symbol, differs fundamentally from the conventional notational mode of note-by-note notation so familiar to musicians of the last three and a half centuries. To those literate in the notation symbols of embellishment, the symbols above the modern conventional notation will be legible; they are translatable into the notation on the staff below. In this variation, the large notes on the staff spell out the embellishment symbols. The top line in the treble clef is not a melody in the nineteenth century sense—it is a florid line emerging from the idiom of embellishment. If this 'spelling out' of embellishment figurations is not recognized, the line is mistaken for a 'melody' in the nineteenth century sense and the resulting performance is anachronistic and stylistically out of character. This is an example of the way in which one may move away from the original nature of a structural idiom into one that is alien by misreading the structural idiom represented by a particular style of notation symbols. Such misinterpretation is totally determined by the way one reads a notation and perceives the structural processes of its symbols.

Performance devices also have a notation. For instance, legato and staccato signify, respectively, connection and varying degrees of shortness, notated by a bowed line for legato and, most often, dots for staccato. The *application* of these devices can vary greatly and it is these differences of application that can influence, and even identify, the stylistic performance of a particular cultural era.

The knowledge and skill required for stylistic application of structural compositional devices in the attempt to identify with the performance practices of another era are, of course, essential; the acquisition of historical knowledge pertaining to the laws and practices of a previous age is mandatory. These form the tools with which every serious scholar and professional performer must be equipped. However, before one can concern oneself with stylistic applications, conceptually and in active performance, the fundamental and primary requirement is to identify the structural materials germane to, and processes contained in, the music itself.

Style is inherent in an era's sense and structuring of musical materials and their relationships. In the organizing processes of a musical composition, style resides in the very blood and bones of the musical organism. Instruments, the technological musical media of a period, influence performance techniques but the intrinsic musical style already resides in the creative concepts of the composer which are communicated to others via the notation symbols. Style is not created by performers, although at times superficial and idiosyncratic devices for external effect create a short lived 'tradition' issuing from a charismatic individual or a trendy fashion. In every era, the composer's intentions for performance are embedded intrinsically in the structuring of a composition. *The ideal goal of a performer, as I see it, is to achieve identity of structural performance treatment and performance devices with the structural compositional treatment and devices.* This appears to me as the strongest and the most incontrovertible base from which to spring to consideration of instrument choices and the historical performance practices of various eras and geographical regions. With the music of Johann Sebastian Bach, where density of materials and almost limitless complexities of structure are the norm, this task requires analysis on many levels.

Apropos of my approach to music in general and to Johann Sebastian Bach in particular, I find it irresistible to paraphrase a comment by Albert Einstein [3] regarding his approach to the perception of complex universal systems:

> I want to know how God created this world . . . I want to know His thoughts, the rest are details.

Likewise, I want to know how Bach created his music, the rest are details.

In the analysis on which I am about to embark, you will see charts and analyses of structural devices in a style not previously applied, to my knowledge, either to music or to performance. This is an analytic technique that I created in my student years, and evolved and applied throughout my entire career in my performance and teaching in all media on period and contemporary instruments. A picturesque image of this analytic technique is to represent a music score as a flower, and the analytic technique as separating each unit of this flower, petal by petal and stamen by stamen, in order to learn what makes up the full-blown flower as an entity. Only then can one, with discrimination and understanding, match performance applications to the respective identities and processes of its materials and organization. When all the relationships that blossom from these are understood, the form of an entire composition virtually fashions itself into an intrinsically interrelated, total organism. As a result of this method of analysis, I never tell Bach how his music should be performed; he always tells me.

Anton von Webern, who had a genius for expressing a great deal in compressed musical form, was also gifted in this way verbally, as can be perceived by his following statement [4]:

> Your ears will always lead you right, *but you must know why.*

The converse of this non-mystical, open-eyed, and rational comment by Webern on the need for conscious perception on the part of a musician, is expressed by the distinguished contemporary poet John Ashberry, whose genius is based in words. He has said this about music [5]:

> What I like about music is its ability to be convincing, to carry
> an argument through successfully to the finish, though the
> terms of the argument remain unknown quantities.

In the phrase 'though the terms of the argument remain unknown quantities', Ashberry's attitude towards music expresses a stance that has always been seductive to laymen and professionals alike. It is easier to detain music in the bondage of mystery than to face its concrete nature and to view with open eyes and mind its subtle and multi-encapsulated processes.

I propose to strip music of its various clothes and costumes, and to this end begin by an examination of its fundamental nature, which is present within its building blocks, its DNA so to speak, and its organizing processes. And now, I invite you to shed your notions, for the time being at least, of melody, harmony, specific instrumental sounds, etc., all of which represent music to the minds and sensibilities of Western musical professionals and laymen, and to consider the elements and principles of organization which make the creation of music—any music—and its performance, possible.

The time spans of sound/silence, coexistent and inter-dependent, provide primary materials for organized musical relationships. These indivisible elements are the first to be set down here for examination. Performance issuing from the analysis of the deeper strata of multi-levelled structures and their relationships illuminate the fact that the form, structure, and organizing processes of the composition stand firm, whether performed on harpsichord or piano. The differing sonorities, textures, volumes and even certain performance devices germane to each of these instruments do not affect the fundamental form, structure and organizing processes of the composition. If form and structure are

perceived by the performer in their individual guise and their interrelated organizing processes within the composition, the vision of the
performer truthfully identifies with that of the composer. These organizing processes are independent of the performance medium.

2. Motivic structure in Johann Sebastian Bach's Fugue in B flat major (BWV 866)
Well-Tempered Clavier, Book I

The technical term for such a fugue as this is 'permutation fugue'. 'Permutation' identifies a particular composing process, but it refers solely to
method; it does not identify the abstract concept of form and structuring
that underlies the method from which this type of compositional organization issues.

The fugue in B flat major (BWV 866) is constructed with three
motives; each motive, in fugue form, is termed subject and/or countersubject. 'Subject', employed in its fugal role, denotes a main motive of
the work which appears initially. 'Countersubject' also denotes a main
motive, but it usually follows the initial announcement of the subject,
and subsequently appears with it, often quite consistently, in various
positionings and locations throughout the fugue. In this fugue, the countersubjects are present with every entry of the subject throughout the
fugue; no other material is present in any form. Therefore, the two countersubjects achieve virtually equal status with the subject. The fugue may
be categorized as having a subject with two countersubjects. One may
also classify it as a triple fugue built with three subjects. The terminology is optional. What is important, incontrovertible, and unalterable by
diverse views of formal analysis, is the fact that there are three motives
employed equally throughout the entire fugue.

I view the three motives as comprising a subject, the first entry, and
two countersubjects. The initial motive, the subject, comprises a linear
and harmonic construct built on a time value scheme that doubles on
itself in the second half of the subject. The two motives that follow are
supplemental, confirming and establishing the harmonic format of the
opening motive and fleshing out its rhythmic design. In this fugue, these
two motives are unchanged at each of their entries and are always present simultaneously with every entry of the subject.

These three motives are, to introduce my own term, 'constant'. I apply
this term to a motive that retains its pattern identically throughout its
every entry in fugal form, no matter what its context and surrounding
materials. A constant motive does not undergo any organic change

within its own identifying figure; its unique interior configurations and relationships are not varied. A constant motive may undergo augmentation or diminution (of its time-values, usually doubled in augmentation or halved in diminution), or inversion (of its intervals), or strict retrograde. Such treatments do not alter its structure, just as the application of certain operations to a geometrical object does not alter its structure.

Illustration 3 The fugue as it appears in its full score.

Illustration 3 shows the entire fugue with its three constant motives, containing all their musical elements and clothed in all their individual compositional processes. Illustration 4 focuses on the exposition of the fugue in which the three constant motives are introduced. These appear here in the full regalia of their pitch patterns, their time-values, their harmonic, melodic, and rhythmic configurations, as well as their implicit linear and vertical relationships to each other in all these aspects. They represent a good deal of complexity, and in addition contain an even greater potential for *developing* complexity in the progress of this form when they undergo reorganization within the vertical framework, a process that begins immediately after the exposition of these essential materials. In the view of the twentieth century, and in much theoretical analysis of the nineteenth century, development of materials has been considered to be dependent upon organic development processes that modify the materials as first presented. This approach has often been imposed, indiscriminately, on fugue form. In actuality, the development of complexity can be accomplished by other means. In fugue, the processes of *positioning*—linear, vertical, diagonal—are the active fabricating procedures that form the structural framework and relationships of the fugal composition and ultimately the organic whole.

Illustration 4 The three motives from the fugue exposition.

c.s. – countersubject

Time—duration of time-values

All the charts here are to be read linearly, that is according to each single part, and vertically, according to the three-part structural relationships that become increasingly exposed.

An introduction to the process of stripping the fugue of conventionally understood musical elements in order to perceive its fundamental framework is presented in Chart 1. This chart also depicts, as does illustration 3, the entire score of this fugue. However, it demonstrates solely the durations of the time-values present in this entire composition. There are no performance additions or emendations here; this is precisely what appears in the original score with pitch notation removed so that the durations can be clearly seen. Devoid of pitch and time signature, the base temporal design of each motive can be clearly perceived and, equally importantly, the temporal relationships can be seen in the progress of the whole composition as all three parts operate simultaneously. Shorn of linear pitch design and harmonic framework, the temporal relationships reveal density of processes and unceasing variety, even to the unpractised eye.

It is these temporal relationships that demand both conceptual and technical articulation on the part of the performer. The nineteenth century emphasis on pitch relationships (i.e. melody or melodic line) and the organizing processes that depend on pitch alone, distracts or confuses these multiple temporal relationships. These multiple relationships exist linearly, vertically and diagonally. The perception and understanding of the temporal design of each part and the temporal relationship of part to part to part in all their positionings and directions must be clearly understood before the element of pitch enters into consideration. The great danger of placing pitch movement first, as is usually done, lies in obstructing the perception of the most elemental skeleton of the composition, which is its *temporal* shapes and designs.

Here in Chart 1, the framework of the fugue is clearly visible; no distractions of pitch-dependent processes intrude, with their inevitable implications of musical formulation, such as melody, harmony, and pitch-related counterpoint. This initial phase of analysis forms the foundation for perceiving fundamental materials and relationships in the ultimate planning of a performance. Certainly, volume of sound or texture do not enter into this phase.

Notice that there are no time signature and no bar lines in this chart. Hence the sole clue to temporal dimension within each figuration is a very simple one: those time-values connected with a single line (quavers) have twice the value of those that are connected with double lines (semiquavers). This chart is a theoretical demonstration; it makes a use-

Chart 1 Entire fugue BWV866: linear and vertical time patterns and relationships.

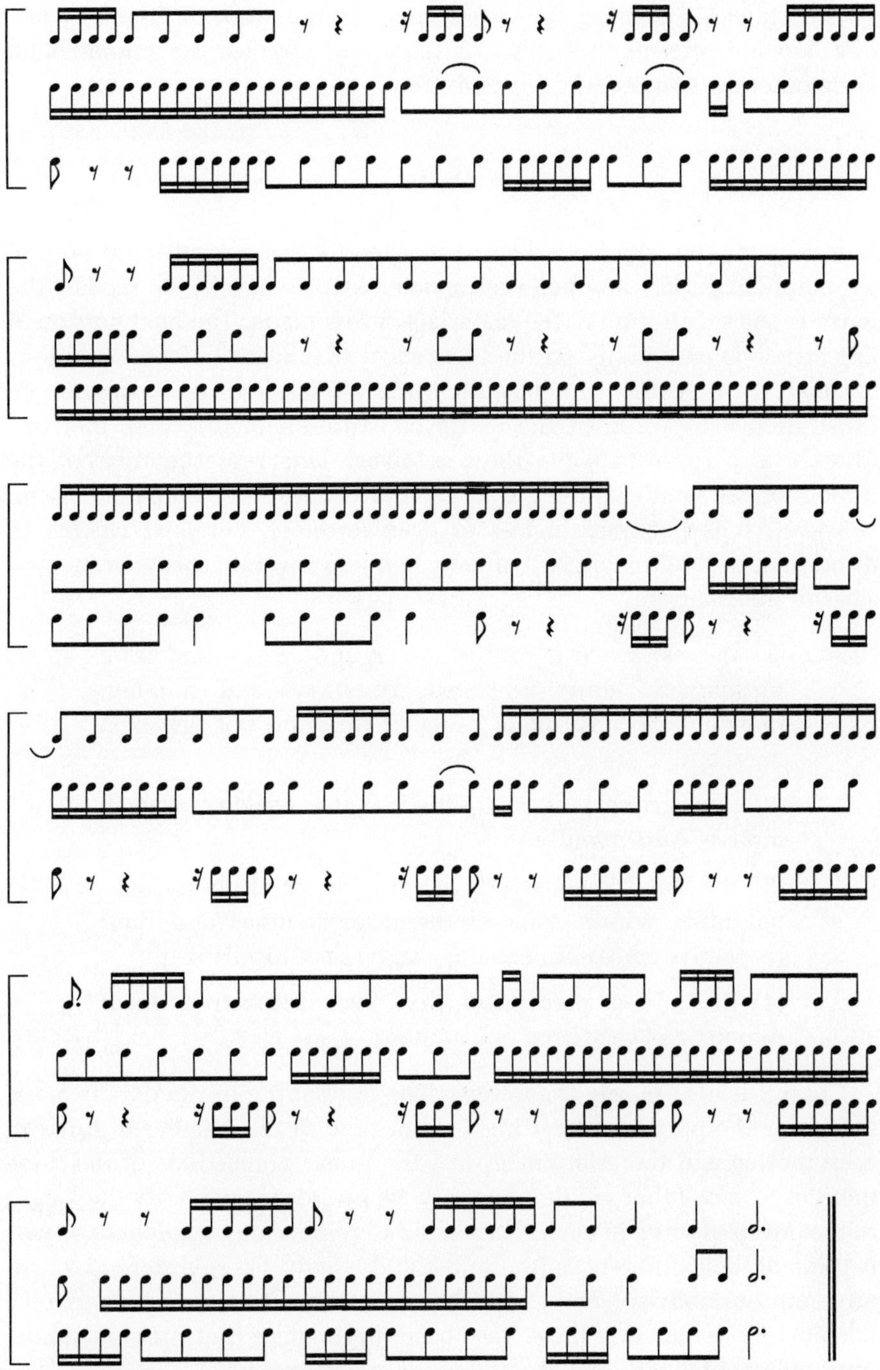

S. – soprano
A. – alto
B. – bass

s. – subject
c.s. – countersubject

ful step towards clearing the visual image in the mind of the many factors that are present in a musical score and clarifies the fundamental
skeleton of the composed designs in time.

Units

Having made the leap from the full score to its transparent framework of
temporal durations, another even greater leap is now to be made—the
move to the substratum of the materials of this fugue. The next portion of
this article is essentially on the theoretical plane, an indispensable area
for analysis, preceding the building-up of the fabric of the composition.
Illustration 4 represented the entire constitution of the three motives.
Chart 2 also represents the three motives. This representation of the
three motives would appear to be devoid of meaning, shape and form.
However, it is not meaningless or even formless, nor is it lacking in
dimension or relationships. It already conveys several aspects of the first
stage of information:

> (1) the *existence* of units—notes and rests—that is, a
> *presence* of units is stated. Durations and *number*,
> inevitable in any set of units in time, do not yet exist
> here, but,
>
> (2) dimension is perceptible in the *length* of each
> motive, Additionally,
>
> (3) the simultaneous vertical coincidence of the individ-
> ual units within each of the three motives and their
> respective linear dimensions can be seen resulting in
>
> (4) the added perception that their respective linear
> dimensions vary from one another.

Chart 3 is also purely theoretical. The numbering of the dots is made
solely for the purpose of getting a clear view of the number of units in
each motive and the relationships of the linear dimensions of the three
motives to each other *via number only*. Its purpose is to clarify the length
of the group of units in each motive and to remove any vagueness or perception of their differences in number and length. These differences usually are unnoticed and are therefore absent from performance
considerations but it is these very basic differences that provide subtle,
but fundamental and crucial, influences in performance.

Chart 4, although still theoretical as far as musical organization and
performance are concerned, represents the containment of the complete

Chart 2 Fugue BMV866: the three motives—units in linear time.

S

1 39

CS¹

1 29

CS²

1 30

Chart 3 Number.

S

1 2 3 4 5 6 7 8 9 10 11 12 13 14 15 16 17 18 19 20 21 22 23 24 25 26 27 28 29 30 31 32 33 34 35 36 37 38 39
....

CS₁

1 2 3 4 5 6 7 8 9 10 11 12 13 14 15 16 17 18 19 20 21 22 23 24 25 26 27 28 29
....

CS²

1 2 3 4 5 6 7 8 9 10 11 12 13 14 15 16 17 18 19 20 21 22 23 24 25 26 27 28 29 30
....

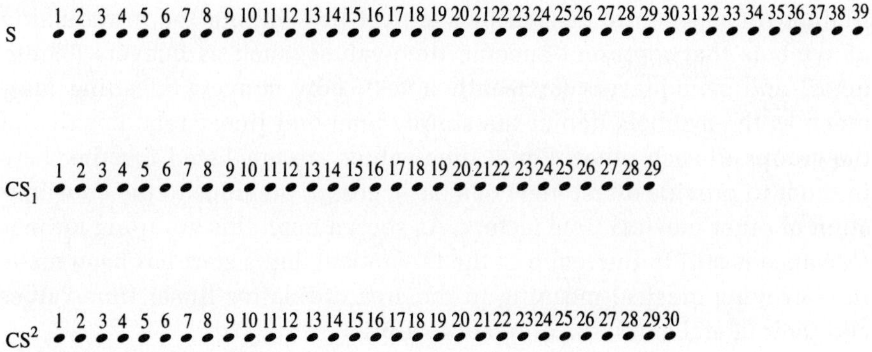

S-subject; CS-counter subject; The units here represent the rests in the subject and CS2, and the ties in CS1, as well as pitch tones.
No rests are present in CS1.

Chart 4 Grouping of linear limits.

S ....

CS₁ ....

CS₁₁ ....

number of units within each motive so that *a beginning and an end* of
the motive is clearly visible; each motive is no longer seen as a vaguely
related number of units. However, no meaningful relationship between
the three motives is yet shown. The bracketing of the outside limits to
the number of notes in each motive conveys the perception of the setting
of a determinate number of notes within a determinate group. This paves
the way for perceiving the relationship of group to group.

Linear time-values and relationships

Chart 5 shows groupings according to: the successive grouping of equal
time-values within their motive's boundaries; related *linear* time-values
of the units of each motive: and lateral and vertical relationships of each
group. This constitutes the first stage of forming an ordered shape.
Neither time signature nor barlines are yet present, but the introduction
of symbols that represent specific time-values, such as quavers (eighth
notes) and semiquavers (sixteenth notes), now conveys meaning inas-
much as the symbols depict the simple temporal linear relationships of
the groups to each other. Equal time-values are bracketed together here
in order to provide a first view of a basic group, previous to the consideration of other musical time factors. As shown here, this grouping for performance is still in the realm of the theoretical, but a start has been made
in conveying musical meaning in the area of relative linear time-values
and their lateral and vertical relationships.

Each group depicted here is equivalent to a structural musical cell.

Chart 5 Groupings according to relative time values, with
equivalent units in determinate groups within the whole group.

Each cell in each part contains its own structures creating relationships within its own organism. According to this concept I arrive at the selection of cells and, having identified them, the formation of groups emerges. The analyses of the internal structures of each cell, which in linear analysis forms a group, and the relationships of group to group, linearly, vertically, and possibly, diagonally develops as I pursue the building blocks and internal processes embodied in the movement in time of pitch tones, time values and more complex designs compounded from these elements.

Groupings according to barlines and time signature

So far simple linear time relationships have been considered—in the form of the quaver and the semiquaver in relation to each other, linearly and vertically. Chart 6 introduces the time signature and barlines; a barline is a counter presence of a time signature. These constitute a mammoth step in the evolution of music. They are introduced here as a normal step from the temporal substratum of the motives to an initial stage in shaping related time-values into discrete patterns. However, before pattern is distinguishable, the units within each motive must be contained within measured time frames. The time signature indicates the number of beats within each bar, each beat here being reckoned as a crotchet (quarter note). Thus, the ordering of units in Western measured music is accomplished by the compound process of a compound time signature which represents both the number of beats and the time-value of each beat. The barline marks the limit of the number of beats from measure to measure. No pitch is yet shown here, but the linear temporal organization of each motive is now apparent. Notice the silences indicated by the rests that occur at the opening of the subject and within the

Chart 6 Groupings according to time signature and barlines. (Barlines are implicit in the time signature.)

second countersubject. These silences are not vacuums; they play a part in linear and vertical relationships equal to those time-values activated in sound. Note the inclusion of the rests within the bracketing of the entire motive in the subject and second countersubject. Heretofore, a primitive, linear, time relationship was perceived in the three motives based on groups of equal time-values. Here, perception of the *shape* of each motive is based *solely* on the measured frame provided by each barline; the motives are now structured within a measured time frame.

Metre

The regularized measuring of music was the primary temporal departure in which Western music separated itself from that of the East. It also led to the creation of harmonic and contrapuntal concepts and techniques. Regularized measuring of pitch patterns in time was equally responsible for bringing music out of the Western medieval world into the modern world. The compound time system of two inseparable entities, the time signature and the barline, establishes a system of measurement with which time-values and groups are mensurated. Regularized measurement—mensuration—made possible the fruitful growth of harmonic form and structures. As polyphony becomes increasingly complex, it leads to the necessity for rhythmic coordination that, in turn, enforces a heightened awareness of simultaneous shapes and sound. The barline and time signature become a necessity, for they insert order in the linear and vertical progress of simultaneous events by creating distinctive temporal patterns and relationships. An inevitable consequence of forming ordered temporal patterns is the emergence of strong/weak time relationships. Downbeat/upbeat relationships, representing strong/weak, provide the first stage of metre from which rhythm is formed.

Based on the measure-to-measure framework created by the barline, which created one type of musical spatial limit, the presence of a time signature as in Chart 6, clearly defines time-values with a prescribed time span. The particular time signature here specifies three crotchet beats to the measure. Besides the arithmetical specification, the time signature also carries a *shaping* presence. In Chart 6, the number three specifies the *number* of crotchets in the measure; number and barline together produce strong/weak metric relationships. With the addition of the barline and time signature new shapes emerge, quite different from those perceived in the primitive linear grouping of equal time-values, as in Chart 5. The latter shapes were useful as a preliminary, clarifying step.

The bracketings in Chart 6 now conform to the group of *beats* within the barline, rather than to the group of equal time-values, and the groups thus formed are made up of *differing* time-values. These groupings do not contradict those of the earlier chart; those had their own validity. These take a different shape because they are perceived in a new category, introduced here for the first time.

From this synthesis of time-signature/barline, another patterning emerges. Here is a first emergence of *groupings*, which form patterns based in a compositional process that is dependent on time signature/barline synthesis. No musical conclusions can yet be drawn. A great many additional factors await unveiling in single categories: their compound entities form still later stages for analysis and evaluation. Structural perception of all the musical factors and operations as well as selective planning for shaping in performance is a long way off.

Groupings, according to beats within the bar

The next stage of subdivision is shown in Chart 7, with groupings according to *each* beat *within* the bar. These groupings pertain solely to the time-value instruction of the time signature. No other composing factor is active here, but the continuous subdivision is useful in perceiving the temporal *cells* of each motive; these issue from units of specified duration. Durational relationships result from these specified units of duration within each barline, to be considered later. This earliest stage of analysis via time-value groupings already reveals a counterpoint of shaping present in the vertical relationships of the three motives. In Chart 7 the primary groupings shown in each part have been assigned according to the quarter beats within each bar. This highlights the variety present in each part on identical beats, viz. bar 1—quavers in the subject, a pattern of semi-quavers and quavers in countersubject 1; a

Chart 7 Groupings according to beats within the bar.

pattern of rests and semi-quavers in countersubject 2. Continued diversity of time values occur on each beat throughout the statement of each motive and, in appearing simultaneously throughout the fugue as they do, produce a continual multiplicity of groupings, i.e. shapes. In this very first stage, in the skeletal condition of time-values alone, the diversity of shaping is prodigious. The need for delineating shapes in performance is clearly revealed. This is, however, only the beginning. Many other musical factors must enter into final decisions as to shaping.

3. Audible events in time—music performance

Since music is an art that is based in audible events in time, one of the *performer's* chief tools must be related to and be capable of depicting *events in time*. So far, I have addressed only a few compositional devices for depicting events in time. But performance also possesses such tools. These are not so largely dependent on volume levels of loud and soft as is conventionally thought; they are primarily dependent on, and achieved through, *durational* structuring.

Durational structuring in composition is based on two simple elements: long and short, and their variable relationships. The long/short device for time spans in performance is achieved by means of connection and disconnection of time-values. In the idiom of musical performance these devices are identified by the familiar terms *legato, staccato, detached, tenuto*, which have been employed for, roughly, the last three centuries. In my view, the ultimate placement, groupings and patterns of long/short in performance depend not on simple arithmetical calculation or on 'feeling,' but rather on all the features that make up the structural totality of the musical organism. This involves diverse factors, such as harmonic functioning and linear rhythmic structures and relationships. These can be reckoned, judged, and selected in final choices only *after* the original DNA of each motive is discovered, and after passing through all the varied processes of formal structuring present in the entire composition.

Figure and pattern

The temporal time-value relationships of the figurative aspects of each motive can be easily perceived in the time-value relationships of the motive itself, but the ultimate patterning of the figure must be fashioned to fit all the progressive transformations of the total context in which it is placed throughout the entire composition. When these are fully perceived and understood, it will be found that these strongly influence the shaping of each motive. More often than not, the original perception of the motive considered solely within the limits of its own existence is thereby changed. Transformations germinate from the counterpoint of temporal patterns in all the parts, but it will be seen subsequently that the linear counterpoint and harmonic development play an integral part in the ultimate performance patterning (phrasing) of individual motives.

The depiction in performance of long/short durational patterns inherent in the musical structuring is dependent on legato (connection) and non-legato or staccato (disconnection). Identifications for varied degrees of disconnection are in an abject state of poverty. Musicians have not developed a refined language or table of symbols for signifying diversified durations of disconnection. Non-legato is a vague term meaning not connected, but the degree of disconnection is unspecified. Staccato is a generic term meaning short in duration. Non-legato and staccato derive from the same basic device, that is, disconnection. Non-legato implies minimal disconnection. Staccato is essentially unspecified in degree (except for an occasional direction of staccatissimo meaning very short, or a wedge-shaped symbol signifying, more often than not, a short, sharp staccato). Aside from these infrequent indications, staccato is rarely accorded variations of durations, either in indications by composers and editors or in actual performance. Yet I find that varieties of degrees of detached touches and treatments for the fulfilment of the myriad time and rhythmic figurations and the compound linear and vertical relationships these create are indispensable. Moreover, varieties of degrees of disconnection are one of the delights in producing and in hearing sounds, whether in primitive or in highly complex music. This aspect of performance is one of the more subtle and indispensable requirements for integrating performance with densely structured music; it is also one of the most demanding, technically.

Varied degrees of detached touch articulate, as they deserve to be, the elements of varied time-values and their relationships as well as the processes that these temporal relationships undergo. In densely composed music, articulated varieties of touch produce unambiguous defining devices for multilevelled contrapuntal structures. The performer's primary tools for depicting such figurative events and relationships in musical time are a true legato and total control over the varying lengths and degrees of shortness. Volume of sound—degrees of loud and soft—do not depict such events in musical time; volume treatments reside in another area of considerations about sound relationships in performance.

Performance devices

Selection of the performance devices that are applicable to the materials of a composition begins on a similar substratum level as does analysis of composed materials. Here too, the initial stages are theoretical. Chart 8 is equivalent in its fundamental state to Chart 2. The dots represent the constant motives of the fugue, as seen in the previous illustrations, contained within the bracketings, which indicate the beginning and end of each motive. The symbols seen above the dots are generic *performance* devices. There are three sets of symbols here, the long bowed line, denoting legato, from beginning to end of each motive; the dots, denoting staccato, above each note of the motive; the short lines, denoting non-legato, appearing above the dots. Each of these symbols represents a specific generic device for shaping sound durations. Although staccato and non-legato issue from the same basic genre of disconnection, the minimal degree of disconnection that is implied by non-legato sets it apart from the still more vague appellation of staccato which is always understood as shorter in duration than non-legato.

The emergence of diversity of groupings, created by time values and matching performance devices that are based in equally primary stages of sound/silence, produces unequivocal time-value patterns. These are based in the very material of the composer's musical ideas/motives. Connections and disconnections are the elements in varied durations of sound/silence that make it possible in performance to shape units of sounds into *relationships* that create *patterns*. From the two basic cells of legato (connection) and staccato (disconnection), Chart 9 shows the

Chart 8 Generic performance devices that represent durations of sound in time: legato, staccato, and non-legato.

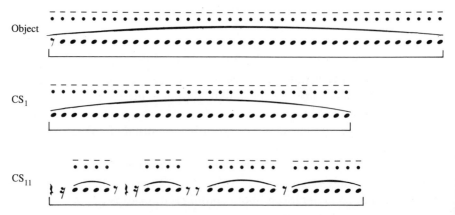

Chart 9 Performance devices, according to relative time-values, conforming with equivalent time units within groups.

initial stage of applying connection and disconnection to the temporal relationships of each motive.

These shapings are based solely on the groupings of *equal* time-values; no other factor is yet being considered. However, the significance of these applications lies in the fact that in performance the original short-valued notes are treated precisely as *short* values, that is, staccato, and the longer-valued notes are treated in performance precisely as longer valued figures by way of legato. In this way, the performance devices match identically the temporal compositional devices of each motive. In this very first stage, intrinsic patterns already become visible. Many other factors enter into the fabrication of a motive's total organism and therefore, other applications besides these two basic cell devices of performance must be considered equally importantly and according to the operation of other structural means, for example metric, harmonic and linear pitch (melodic) progression. These are to be discovered and analysed in later stages. Further, their analyses lead to compound considerations embracing all of these factors.

Linear variants and relationships in groupings according to barline/time signature, beats, and time-values

Below the units of each motive in Chart 10 is a bracket. These identify with the groupings of equal time-values, as in Chart 9. The original bracketing of the groupings in the very first stage (showing the equal time-values) is retained on the chart to keep clear the stages traversed up to this point.

The performance devices are diversified. Notice that in (a) the designs

Chart 10 Linear variants and relationships: (a) Linear time-values, b) barline/time signature, (c) compound demands beat-by-beat measurement and long/short relationships, and (d) performance inversion in subgroups creating patterns linearly and horizontally.

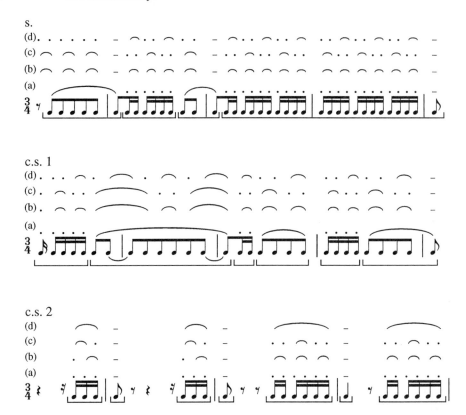

are derived from the original temporal structuring of long/short, the quaver (long) being halved by the semiquaver (short). The performance devices are based on the linear time-values in conformity with the composed pattern long/short; therefore, for performance the quavers are viewed as legato, the semiquavers as staccato. In (b) the designs are based on the barline and the beat-by-beat pattern indicated in the 3/4 time signature; therefore, ⏜ over pairs of semiquavers in bars 2, 3, and 4 is based on the pairing of the strong/weak relationship of downbeat/ upbeat present in the temporal downbeat demands, characteristic of beats 1, 2, and 3 in a time frame of 3/4, the crotchets in the measure, and in the relation of the first two semiquavers to the second two semiquavers within each crotchet. (c) and (d) indicate inversion possibilities of (a) and (b).

In c.s. 1 (b, c, d) the design is also determined by the strong/weak temporal relations of the crotchets in each measure, due to the 3/4 time signature, but the tie in the third beat of bars two and three pulls the weak third beat into the strong first beat of these bars silently, extending the time-span of one pitch through the barline; therefore allowance is made here for this extended time-span created by the tie by extending the phrase shape into the succeeding bar.

Inversions

The possibility of structuring in performance includes many devices employed in compositional structuring. Inversion of durational treatments is one such structural performance device. A two-note slur or long legato design may be inverted to a disconnected design. Such inversion cannot be made indiscriminately or for fancy effects. There are multiple factors that may be at work in the original structure that call forth inversions of this nature. All factors must be considered before such performance inversions are introduced into a motivic design. The patternings of ▬ and of ·· in (c) and (d) in each of the three motives constitute inverted treatments of (a) and (b), as well as of each other. Other inversions are also present, for instance: the (d) design of c.s. 2 constitutes an *inversion* of the three designs in (a), (b), and (c) in that it is *unvaried* in its legato treatment of the group whereas (a), (b), and (c) are varied in long/short designs. The crucial downbeat of each bar in c.s. 2 cannot be varied for reasons compounded by harmonic and rhythmic factors, to be considered later.

Time relationships—linear and vertical patterns—entire fugue

Chart 11 relates precisely to Chart 1, which contains the entire fugue. Both are stripped of all pitch, harmony, and rhythmic measurements such as those that emanate from the time signature and barlines, but the linear time-values and their relationships are delineated here. The crucial consequence of this delineation is a clear perception of the linear and vertical patterns and their relationships throughout the fugue. Already, even at this initial stage, it is clear that the linear patterns and the vertical patterns engender compound structures of connections and disconnections, that is, in terms of performance, legato and staccato. This patterning, made strictly according to the temporal values of the music score, projects solely the raw initial stage of the linear and vertical

Chart 11 Time relationships, linear and vertical patterns; entire fugue (first stage).

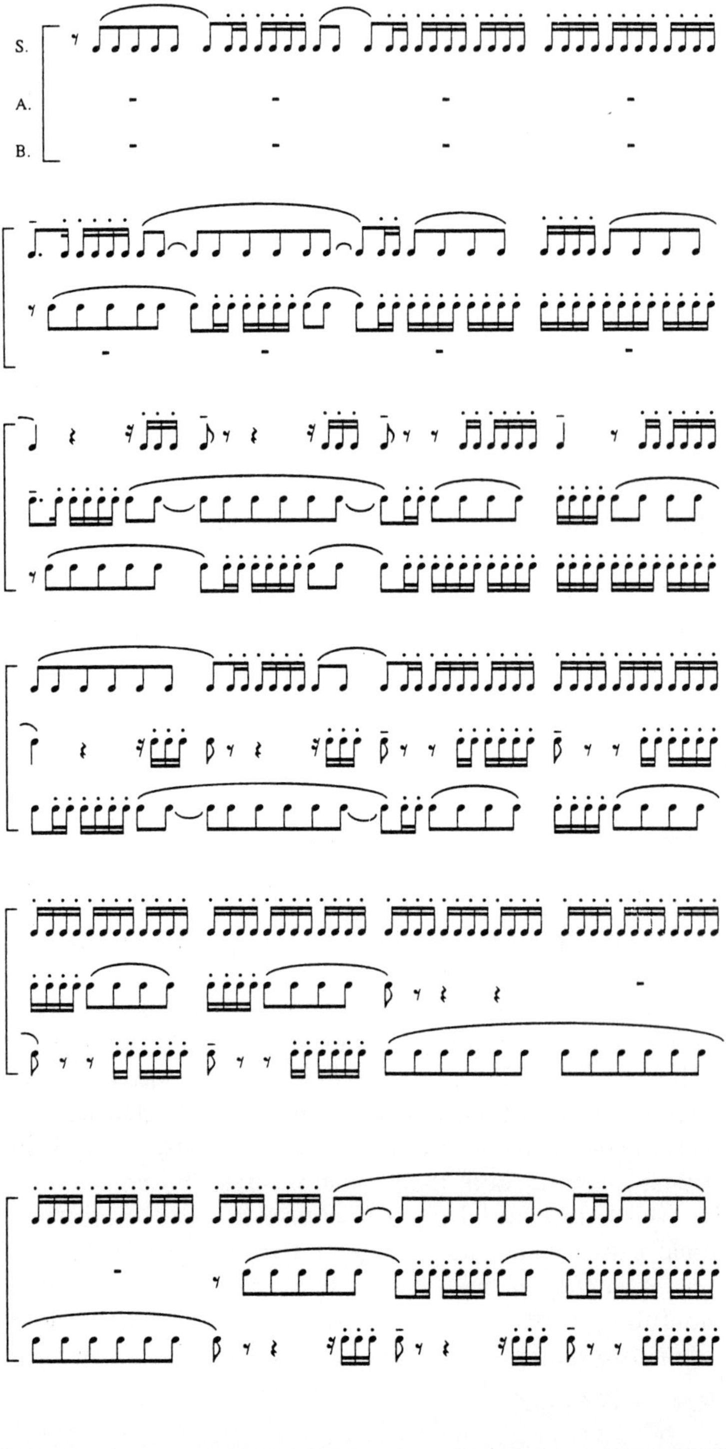

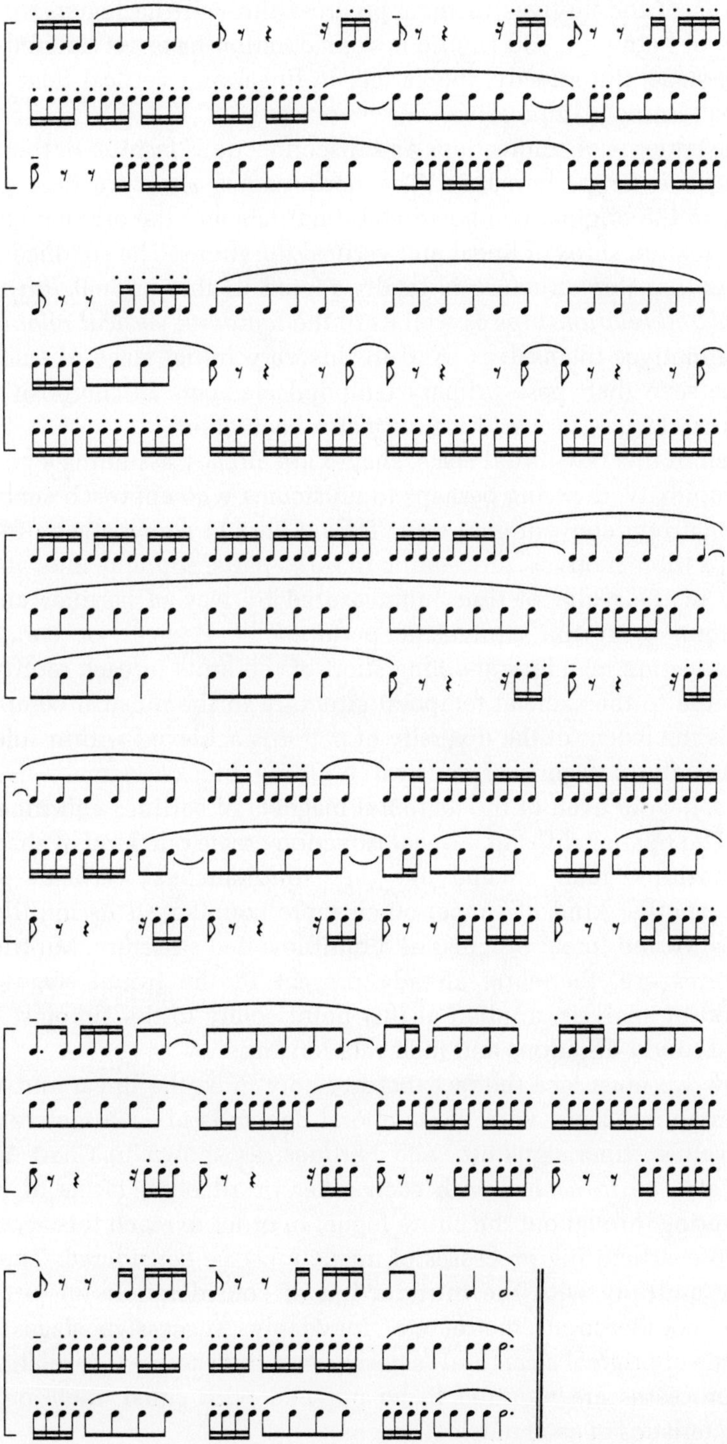

patterns of the motives in their progress through the fugue. No refinements of harmonic and rhythmic consideration have yet entered in this initial stage. But already, the variety in linear and vertical relationships and patterning is impressive.

The patterns of connections and disconnections fashion in themselves a structure for performance. The performance structure here pertains solely to the original compositional durations and the original compositional relationships of linear and vertical durations. The significant point here is that this structure is totally rooted in the original *linear time-values and relationships* as well as in the *temporal vertical relationships* of the motives themselves. And in this very initial stage of analysis, it can be seen that these primary temporal elements in composition and temporal devices for performance match each other.

Even in this raw initial stage, shapes are already assuming a good deal of complexity, dizzying perhaps to musicians who approach such music in the current conventional way. Thus, Chart 11 shows the entire fugue with its three motives, proceeding in three parts, soprano, alto, and bass, solely in the realm of time, unmeasured by way of barlines and time signature, with the equivalent performance devices of connection/disconnection matching the long/short of the units in each motive. This reduction to the skeletal temporal structure in the musical composition reveals the extent of the diversity of patterns achieved within solely one fundamental category of musical building blocks, namely the time-values, devoid even of the temporal magnets of barlines and time signature. The relationships in a linear direction create one kind of shape. The relationships that emerge in the simultaneous vertical patterns create another kind of shape, much more complex in its multilevelled verticality and linear progress as a multilevelled structure. Multilevelled structures are, therefore, already present in the initial stage of this method of analysis, applied at this point solely to the simplest dimension of sound-durations and their relationships.

Now we must face the fact that not only must the linear and vertical shapes be in accord with the temporal demands of each motive's given time-values, time signature, and barlines, as shown in Chart 11, they must also be *in accord* with each other in all other facets of musical structuring throughout the entire fugue. In order to reach this accord, the multiple structuring processes of music must be considered. These processes multiply with the introduction of considerations of pitch, harmony, counterpoint, metre, etc. Inevitably, successive stages in the analysis of original structural materials and their respective relationships and processes are required to be pursued with equal attention to the characteristics of each musical process.

The shaping of time-values and their relationships is comparatively straightforward. This stage forms only the beginning, even in relation to the temporal aspect of music; the subtle relationships of rhythm, for instance, cannot be considered until other cells, functions, and processes of musical materials are considered. Rhythm depends on and emerges from all these other factors. Ultimately, the individual and compound relationships of the varied aspects of the nature and function of the musical materials must be examined separately and then related to each other. All are equally intrinsic to performance structuring and all play virtually an equally important part in the final selection of performance devices that match and identify with the totality of the compositional structure. No single element may be ignored or passed over; if so the structure of the composition is weakened to the degree that certain elements are absent. Distortions and anachronisms are born in this realm.

Harmonic structure

A glimpse into later stages is provided in Chart 12. Here is a sample of an initial stage in dealing with the harmonic skeletal structure of the opening subject. The bracketing of the harmonic traversal of the subject, from the subject's opening and closing notes as well as its harmonic subsections, serves to clarify the harmonic shaping of the subject. The bracketing of the harmonic subsections serves to clarify the inner harmonic shaping. The structural perception of these inner and outer

Chart 12 Harmonic structure.

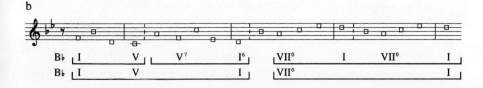

frames etches the image of the shape(s) of the subject that issue from its
harmonic fabric.

In (a) the inner structure of the subject is analysed according to its
inner harmonic skeletal divisions. In (b), all temporal factors are stripped
from the essential pitch tones that form the framework of the harmonic
process in this motive. It will be seen that the overall thrust of the harmonic shape for this motive is based in tonic-dominant-tonic—I–V–I.
The breakdown into subdivisions, in both (b) and (a) show how the
larger harmonic framework of I–V–I is accomplished in its progress
through four bars. In order to understand the deeper harmonic relationships, it has been necessary to depict the basic steps of the harmony and
their function within the four-bar motive.

But there is a good deal more than meets the eye. The harmonic relationship of bar 2 to bar 1 functions as a departure from the tonic, but it
also functions as a harmonic bridge to the coda of the subject, which
appears in bars 3 and 4. Nestling in this four-measure motive is another
harmonic design. The design of bar 2, V–I, in relation to bar 1, I–V, is
inverted. Bars 3 and 4, VII^0, are in inversion to bars 1 and 2, I–V. The
four-bar subject traverses a fully executed harmonic statement, depicting
the key of the composition by way of the conventional harmonic
cadence, I–V–VII^0–I.

This harmonic analysis demonstrates an instance of a basic harmonic
configuration. Its shape and progress must eventually generate significant influences on performance, choices of performance structuring and
of performance devices intrinsically suited to the original harmonic
shaping and progress.

Melody

Melody, or melodic line, is another essential element in our Western
music. A consecutive succession of pitch tones in a linear design is the
usual configuration given to the Western image of melody. Musicians of
the nineteenth and twentieth centuries whose education is primarily
based in nineteenth century values of melody and harmony, still the
mainstay of the structural approach today for most performers, tend to
latch mainly onto the linear aspect of melody in what I consider to be
the initial stage of musical perception of a composition. A melody
depends on factors other than linear factors in order to attain an organic
life of its own. Unlike time-values, which on the strength of their symbols depict precise arithmetical relationships that are distinct and recognizable, melody must depend not soley on linear pitch patterns which,

although in themselves are distinct and recognizable, do not establish a melody or even a melodic line. Multiple factors such as time-values, time signatures, rhythm (which is more ambiguous) and the harmonic inner relationships and outer framework are indispensable in the creation of a melodic figure. Melody, cannot exist independently of the primary element of time-values and their relationships and, in the developed idiom of Western music, of a harmonic framework, explicit or implicit.

Chart 13A shows the structure of the opening subject in terms of its time-values and the relationships that the time-values and beats form within the bar. The equal time-values are bracketed.

The brackets depicted in (A)1a represent the first stage of analysis, demonstrated in previous Charts 5, 6, and 7. This first stage is as contributory to the composite factors present in the ultimate perception and phrasing of melody as it is to the temporal facets of a motive or figuration. Based on this initial perception of groups within beats containing solely equal time-values are musical shapes, (A)1b, 2b, the shorter and longer slurs. These slurs do not represent performance treatment *per se*; they indicate the mental image of shaping that emerges from the perception of this group.

Similar treatment is applied to the longer groups representing an

Chart 13A Melody: (a) time values, (b) groupings and (c) their contrasted and expanded relationships.

Chart 13B Performance devices. Variants of inner articulation based solely on consecutive equal time values, barlines, time signature, first beats of each bar. Variants include inversion of performance devices.

entire sub-phrase spanning a full measure rather than the sub-phrase of single beats. These are first bracketed as in (A)2a, also seen in Charts 5, 6, and 7. The musical shape, (A)2b, emerges from the perception of the group.

(A)3a provides the initial stage of relationships of short-valued notes (in measures 3 and 4) to the longer-valued notes (in measures 1 and 2). (A)3b indicates the full length of the two motivic phrases that constitute the opening subject.

A deeper level of scrutiny applied to this subject illuminates the fact that there are (1) two motivic temporal elements here, the quaver and the semiquaver; additionally (2) there are two ways of employing these two elements, connection and disconnection; and (3) there are two performing treatments of these applied to the two motivic figurations of this subject—legato and detached. Moreover, it can be perceived that bar 2 contains nothing new in the way of a motive, for bar 2 simply reiterates bar 1, but it amplifies measure 1 in two ways: harmonically, in reiterating measure 1 in a sequence one tone above bar 1; and temporally—it introduces the semiquaver time element. These semiquavers serve two functions—firstly, they constitute the entire temporal figuration of bars 3 and 4 thereby forming an additional motivic figure in the subject. Secondly, they introduce a new temporal element and combine in a relationship of rhythmic diminution (halved time-values) with the quavers of bars 1 and 2. To recognize these intrinsic elements and relationships of time-values, their role in forming motives, and their timespans within the four-measure subject is to perceive a fundamental treatment of these materials that will affect and contribute to the intuitive perception of the entire fugue, as well as to the conscious understanding of the myriad of linear and vertical designs that these relationships form.

Chart 13(B) shows varied articulation choices for performance treatment that emerge from the composite considerations of time-value relationships, time signature, and motivic timespans. But these indications still provide only the raw materials of performance choices based on simple temporal values. More aspects of compositional structuring must now be examined, such as the harmonic configurations and the metric consequences of temporal and harmonic configurations, etc. Melodic shaping is dependent on these for ultimate choices for phrasing and articulation in performance.

Conclusion

The consideration of choices for the performance treatment of melody necessitates the combining of musical elements. Chart 13(A) demonstrates possibilities of inner and outer shaping limits for the motive of the opening subject. These shapings (a term I prefer to articulations when referring solely to melodic line) constitute a mental image that is based on the time-values, barline, time signature, and first beats. First beats contribute metre and rhythm to the figure within the bar and to the full length of the entire motive that here extends through four bars. In Chart 13, a metric pattern has been introduced. The significance of such analysis as in Chart 13(A and B) lies in demonstrating the multiple possibilities of shaping inherent in the *rhythm* of musical figurations as well as the elements of time-values.

The articulation symbols in Chart 13(B) emerge from a dependence of the linear succession of pitch tones upon their placement in the metric position within each measure as well as their primary time-value relationships. There is still no involvement here with the harmonic framework and its internal progressions. This illustration contains the consideration of subdivisions pertaining only to durations, their relationships and a minimal level of metre, owing to the presence of a time signature and barlines. A number of theoretical choices for performance shaping become visible in this initial stage of melodic analysis.

To demonstrate a final choice based on all structural factors—temporal, rhythmic, harmonic, melodic—I shall make a great leap over the requisite additional levels of analysis. Chart 14 shows one valid final shaping for the opening subject. These articulations emerge from the successive stages of analysis of the temporal, melodic and harmonic elements. The legato (connected) treatment for the first phrase in equal quavers is based on the temporal element—long values (quavers) in relation to the short values (semiquavers) in the following phrase (measure 2). The harmonic progression pulls the phrase of measure 1 into the first downbeat of measure 2; here both the harmonic and temporal are the decisive factors. The shaping of the melodic aspect is ruled here by the harmonic and temporal, as are the following phrases (groupings) in measures 2, 3, and 4. The groupings are easily distinguished by the harmonic progressions of this subject, indicated in Chart 14, below the staff. The inevitable magnetic pull of harmonic relationships combine with that of

Chart 14 Performance. A sample of ultimate shaping and articulation based in and emerging from the integration of fundamental temporal, harmonic and melodic elements, processes, and relationships.

I⌐   ⌐V⌐L   ⌐V⁷   I⌐L ⌐VII°   I   VII°   I⌐

temporal relationships of weak/strong-upbeat/downbeat. These dictate
the outer limits and the metre of the melody. When the three basic elements of music—time (temporal relationships), harmony (simultaneous
vertical constructs), and melodic line (linear pitch relationships)—meet
together in perfect agreement, the phrase shape and the articulations
within the phrase identify with the original compositional elements and
relationships. The consideration of counterpoint and its procedures
throughout the fugue constitutes another type of micro-analysis, equally
essential to conclusive choices as to performance decisions about phrase
shapes and articulations, dynamics, etc. In the analysis given here I was
concerned with the initial stages of composed music, structuring of the
primary elements of music—time and pitch.

Based on so strong a foundation as the materials and the processes
that they undergo, the performance treatment carries conviction. It will
be seen that the 'feeling' response is spontaneously fulfilled because all
the incontrovertible strong points of the structure are fulfilled. Chart 14
presents a valid structuring treatment in performance, but this is not the
only one that is valid. The compositional framework and processes stand
firm, but within existing varied choices. The ultimate decision regarding
fine points in articulations is the choice of each individual performer;
therefore this is not a robotic or mechanically contrived system producing copies equivalent to the duplications of a photocopy machine.

Individuality, the hallmark of great art (when not idiosyncratic), is
released as a consequence of the perception of the underlying processes
of the materials in the music, leading to a greater play of imagination and
personal feeling in choices. Such a release is inevitable because a very
wide range of structures and their attendant shapes and articulations
emerge clearly revealed. Failing analysis in these stages and varied
structural levels, some choices are bound to be missed with the result
that the composer's thoughts are not fully represented and the performance is the poorer for it.

In this necessarily extremely abbreviated exposition of primary musical elements, the structuring of composing elements, and the relationship of primary performance elements and their structuring, only the
most sparing examples could be given. Stage by stage this analytical
technique, which is founded in the most fundamental musical elements
and processes, leads from cell to functions to relationships. These
embrace the inherent structural rules and processes that not only produce the specific configurations of a composition, but also represent the
formations of particularized forms in diverse eras in Western culture. All
of these stages reveal devices for *performance* that produce complemen-
tarity of performance structuring with compositional structuring. The

ultimate result is a multilevelled, organically integrated structure combining analytic perception and expanded intuitional responses in performance.

My aim is to identify with the composer's compositional materials and thought processes. The rest, as Einstein says, are details.

References

1. Excerpt from the synopsis of a lecture given by Professor Henk Barendregt of Nijmegen University at my *Symposium on Structure* in August 1990 at the Universidad Internacional Menedez Pelayo, in Santander, Spain.
2. Debussy, Claude (1991). *Music Quotations* (ed. Derek Watson), p. 5. Chambers, Edinburgh.
3. Clark, Ronald W. (1984). *Einstein: The Life and Times, an Illustrated Biography*, p. 19. Abrams, New York.
4. von Webern, Anton (1991). *Music Quotations* (ed. Derek Watson), p. 99. Chambers, Edinburgh.
5. Ashberry, John (1976). *New York Times Magazine*, 23 May.

ROSALYN TURECK

Born in Chicago, first American generation of Turkish-Russian lineage. Musical studies with Sophia Brilliant-Liven, assistant teacher to Anton Rubinstein, Jan Chiapusso, Dutch-Italian pianist and Bach scholar and Olga Samaroff at Juilliard Graduate School. Debut, piano recital aged 9. Specialized studies of Bach's music, begun at age 14, included organ, harpsichord, clavichord as well as the piano and nineteenth and twentieth century repertoire. Awarded numerous honorary degrees including Oxford University. Conducting engagements of international orchestras include New York Philharmonic Orchestra: the Philharmonia, London; Scottish National Orchestra, Edinburgh, Glasgow; National Orchestra, Washington DC; St. Louis Symphony; Tureck Bach Players, London and New York; Madrid Symphony Orchestra; Kol Israel Symphony Orchestra, Tel Aviv, Jerusalem; Collegium Musicuum, Copenhagen, etc. Visiting Professor: Princeton University, Washington University, St Louis USA; Univ. of California, San Diego, Los Angles, Santa Barbara; Yale University; St Hilda's College, Oxford, Hon. Life Fellow; lectures, Master classes at Oxford University etc. Creator and Director, Composers of Today, Inc., and the Tureck Bach Research Foundation, Oxford. Since the 1940s has formed long-standing friendships with scientists Isidor Rabi, Harold Urey, Niels Bohr and Aage Bohr and, recently, with Sir Rudolf Peirerls, Mitchell Feigenbaum and Benoit Mandelbrot. Publications, besides numerous articles, include: Urtext, facsimile and Performance editions of J. S. Bach (Schirmer), and 'Introduction to Performance of Bach' (Oxford University Press). Lecturer's papers and manuscripts are in the Tureck Collection at Boston University; complete recordings and films are in the Tureck Archives, Library for the Performing Arts, Lincoln Center, New York. Forthcoming books: 'A History of Bach Performance', and 'Cells, Functions, Relationships in Musical Composition and Performance'. Recent recordings in extensive programs of J. S. Bach include: 'Goldberg Variations', 'Partitas', 'Chromatic Fantasia and Fugue', etc. A continuing historical retrospective series of CDs of live performances since 1939 contains music of 18th, 19th, and 20th centuries and world premieres. Recent video films: live concert at Teatro Colon, Buenos Aires, August 1992. Forthcoming videos: Live at St Petersburg, Russia, June 1995; Live at Rome, Italy, March 1996.

This lecture represents material in the first chapter of the above-mentioned, forthcoming book bearing the lecture's title.

Discourses

Incremental Decisions in a Complex World　　　*10 November 1995*
Chris Elliott, MA, PhD, CEng, FRAeS

Science and Fine Art　　　*17 November 1995*
David Bomford, MSc, FIIC

Exploring the Universe with the Hubble Space Telescope　　　*24 November 1995*
Alexander Boksenberg, PhD, FRS

Visual Art and the Visual Brain (The Woodhull Lecture)　　　*1 December 1995*
Semir Zeki, FRS

THE ROYAL INSTITUTION

The Royal Institution of Great Britain was founded in 1799 by Benjamin Thompson, Count Rumford. It has occupied the same premises for nearly 200 years and, in that time, a truly astounding series of scientific discoveries has been made within its walls. Rumford himself was an early and effective exponent of energy conservation. Thomas Young established the wave theory of light; Humphry Davy isolated the first alkali and alkaline earth metals, and invented the miners' lamp; Tyndall explained the flow of glaciers and was the first to measure the absorption and radiation of heat by gases and vapours; Dewar liquefied hydrogen and gave the world the vacuum flask; all who wished to learn the new science of X-ray crystallography that W. H. Bragg and his son had discovered came to the Royal Institution, while W. L. Bragg, a generation later, promoted the application of the same science to the unravelling of the structure of proteins. In the recent past the research concentrated on photochemistry under the leadership of Professor Sir George (now Lord) Porter, while the current focus of the research work is the exploration of the properties of complex materials.

Towering over all else is the work of Michael Faraday, the London bookbinder who became one of the world's greatest scientists. Faraday's discovery of electromagnetic induction laid the foundation of today's electrical industries. His magnetic laboratory, where many of his most important discoveries were made, was restored in 1972 to the form it was known to have had in 1845. A newly created museum, adjacent to the laboratory, houses a unique collection of original apparatus arranged to illustrate the more important aspects of Faraday's immense contribution to the advancement of science in his fifty year's work at the Royal Institution.

Why the Royal Institution is Unique

It provides the only forum in London where non-specialists may meet the leading scientists of our time and hear their latest discoveries explained in everyday language.

It is the only Society that is actively engaged in research, and provides lectures covering all aspects of science and technology, with membership open to all.

It houses the only independent research laboratory in London's West End (and one of the few in Britain)—the Davy Faraday Research Laboratory.

What the Royal Institution Does for Young Scientists

The Royal Institution has an extensive programme of scientific activities designed to inform and inspire young people. This programme includes lectures for primary and secondary school children, sixth form conferences, Computational Science Seminars for sixth-formers and Mathematics Masterclasses for 12-13 year-old children.

What the Royal Institution Offers to its Members

Programmes, each term, of activities including summaries of the Discourses; synopses of the Christmas Lectures and annual Record.

Evening Discourses and an associated exhibition to which guests may be invited.

An annual volume of the *Proceedings of the Royal Institution of Great Britain* containing accounts of Discourses.

Christmas Lectures to which children may be introduced.

Meetings such as the RI Discussion Evenings; Seminars of the Royal Institution Centre for the History of Science and Technology, and other specialist research discussions.

Use of the Libraries and borrowing of the books. The Library is open from 9 a.m. to 9 p.m. on weekdays.

Use of the Conversation Room for social purposes.

Access to the Faraday Laboratory and Museum for themselves and guests. Invitations to debates on matters of current concern, evening parties and lectures marking special scientific occasions.

Royal Institution publications at privileged rates.

Group visits to various scientific, historical, and other institutions of interest.

Evening Discourses

The Evening Discourses have been given regularly since 1826. They cover all aspects of science and technology (with regular ventures into the arts) in a form suitable for the interested layman, and many scientists use them to keep in touch with fields other than their own. An exhibition, on a subject relating to the Discourse, is arranged each evening, and light refreshments are available after the lecture.

Christmas Lectures

Faraday introduced a series of six Christmas Lectures for children in 1826. These are still given annually, but today they reach a much wider audience through television. Titles have included: 'The Languages of Animals' by David Attenborough, 'The Natural History of a Sunbeam' by Sir George Porter, 'The Planets' by Carl Sagan and 'Exploring Music' by Charles Taylor.

The Library

The Library contains over 60,000 volumes, and is particularly strong in long runs of scientific periodicals. It has a fine collection of the history of science and science for the non-specialist. A selection of newspapers and magazines is available in the Conversation Room which is still characteristic of a library of the early nineteenth century.

Schools' Lectures

Extending the policy of bringing science to children, the Royal Institution provides lectures throughout the year for school children of various ages, ranging from primary to sixth-form groups. These lectures, attended by thousands, play a vital part in stimulating an interest in science by means of demonstrations, many of which could not be performed in schools.

Seminars, Masterclasses, and Primary Schools' Lectures

In addition to educational activities within the Royal Institution, there is an expanding external programme of activities which are organized at venues throughout the UK. These include a range of seminars and master-classes in the areas of mathematics, technology and, most recently, computational science. Lectures aimed at the 8-10 year-old age group are also an increasing component of our external activities.

Teachers' Workshops

Lectures to younger children are commonly accompanied by workshops for teachers which aim to explain, illustrate, and amplify the scientific principles demonstrated by the lecture.

Membership of the Royal Institution

Member

The Royal Institution welcomes all who are interested in science, no special scientific qualification being required. By becoming a Member of the Royal Institution an individual not only derives a great deal of personal benefit and enjoyment but also the satisfaction of helping to support the unique contribution made to our society by the Royal Institution.

Family Associate Subscriber

A Member may nominate one member of his or her family residing at the same address, and not being under the age of 16 (there is no upper age limit), to be a Family Associate Subscriber. Family Associate Subscribers can attend the Evening Discourses and other lectures, and use the Libraries.

Associate Subscriber

Any person between the ages of 16 and 27 may be admitted as an Associate Subscriber. Associate Subscribers can attend the Evening Discourses and other lectures, and use the Libraries.

Junior Associate

Any person between the ages of 11 and 15 may be admitted as a Junior Associate. Junior Associates can attend the Christmas Lectures and other functions, and use the Libraries. Junior Associates may take part in educational visits organized during Easter and Summer vacations.

Corporate Subscriber

Companies, firms and other bodies are invited to support the work of the Royal Institution by becoming Corporate Subscribers; such organizations make a very valuable contribution to the income of the Institution and so endorse its value to the community. Two representatives may attend the Evening Discourses and other lectures, and may use the Libraries.

College Corporate Subscriber

Senior educational establishments may become College Corporate Subscribers; this entitles two representatives to attend the Evening Discourses and other lectures, and to use the Libraries.

School Subscriber

Schools and Colleges of Education may become School Subscribers; this entitles two members of staff to attend the Evening Discourses and other lectures, and to use the Libraries.

Membership forms can be obtained from: The Membership Secretary, The Royal Institution, 21 Albemarle Street, London W1X 4BS. Telephone: 0171 409 2992. Fax: 0171 629 3569